ENERGY POLICIES, POLITICS AND PRICES

ENERGY EFFICIENCY

PERFORMANCE, IMPROVEMENT STRATEGIES AND FUTURE DIRECTIONS

ENERGY POLICIES, POLITICS AND PRICES

Additional books in this series can be found on Nova's website under the Series tab.

Additional e-books in this series can be found on Nova's website under the eBooks tab.

ENERGY POLICIES, POLITICS AND PRICES

ENERGY EFFICIENCY

PERFORMANCE, IMPROVEMENT STRATEGIES AND FUTURE DIRECTIONS

JULIAN E. PEREZ
EDITOR

nova
science publishers
New York

Library of Congress Cataloging-in-Publication Data

ISBN: 978-1-53611-040-1

Published by Nova Science Publishers, Inc. † New York

CONTENTS

PREFACE

There are increasing demands on energy supplies as they become an essential requirement to support modern life in most countries. Conventional resources for power generation that depend on fossil fuels become less reliable because of price fluctuations and long-term supply uncertainties. Furthermore, strong environmental awareness coupled with recent efforts to reduce greenhouse gas emissions has encouraged a shift from fossil fuels to renewable power generating resources. This has resulted in a significant investment in renewable energy technologies such as wind, solar, hydro and fuel cells. This book provides new research on performance, improvement strategies and the future directions on energy efficiency.

Chapter 1 – This chapter reports on the research approach adopted in the development of a building energy efficiency assessment tool for a developing country based in Sub-Saharan Africa: Ghana. The inability to solve peak demand constraints by finding a balance between reducing demand and increasing supply has resulted in power rationing coupled with erratic electricity supply in Ghana. Frequently used is the energy standard which is instrumental in enhancing efficient energy design in buildings and overall energy efficient improvements. Internationally, the field of building energy efficiency assessment methods has been characterized by a plethora of research. However, no building energy standards are in place in Ghana and this is coupled with dearth of research focused on these areas in such regions. These complexities and peculiar challenges informed the adoption of an exploratory sequential (mixed method) approach. This approach consisted of a synthesis and analysis of building energy assessment methods with the Delphi Technique which was further combined with both the Delphi-Analytic Hierarchy Process and a simulation based study. Exploring various related research approaches and the undergirding research paradigms, the study discusses the use of the adopted approach against the backdrop of the region's idiosyncratic challenges. Key amongst the strength of the adopted mixed method approach is that it enabled answers to be provided to research problems that would not have been answered using just one single approach. Also, the method built a stronger case for conclusion through convergence and verification of findings and made it possible for a holistic knowledge to be provided, necessary to inform both theory and practice. This approach provides answers which were wide-ranging and broad due to the multiplicity of research questions used. A generic model for the development of a building energy efficiency assessment tool suited for Ghana and similar regions is thus provided.

Chapter 2 – This chapter/research review is a study of the different waste-to-energy technologies (WTE-Ts) developed to date. This study of the technologies is divided into four groups: biological treatment of waste; thermal treatment of waste; incineration of waste; and landfill gas utilisation. Furthermore, integrated solid waste management systems (ISWM-S) with WTE-T are studied and some worldwide examples are provided.

Chapter 3 – Orthogonal frequency division multiple access (OFDMA), the air interface for major and future wireless communication systems, can achieve significant energy saving while satisfying various quality-of-service (QoS) requirements with intelligent management of radio resources such as subcarriers and transmission power. One of the common QoS metrics belongs to proportional fairness, which enforces the achieved data rate proportions among users. Together with this set of nonlinear equality constraints, the problem of energy-efficient resource allocation can be modeled as a mixed-integer fractional program, which is computationally difficult to solve. Separating the subcarrier and power allocation is a prevalently adopted alleviation to reduce complexity while approaching the optimal system performance. Under a pre-determined set of subcarrier allocation, most existing solutions aim to maximize the energy efficiency of OFDMA systems, which belongs to nonlinear fractional program. However, their derived solutions are based on some ideal assumptions such as high channel conditions. When channel conditions degrade, the adherence to the proportionality constraints will no longer be strictly obeyed. In this chapter, our goal is to maximize the energy efficiency of OFDMA systems subject to the constraints on base station transmission power and proportional data rates. The authors first convert the nonlinear fractional problem into a simpler single-variable nonlinear problem via Charnes-Cooper Transformation. By exploring the structure of the new problem, they then develop an iterative algorithm to find the optimal solution of the considered problem while delivering hard guarantee of the proportional data rates. Simulation results reveal that our proposed scheme satisfies the rate proportionality constraint in the strictest sense, as compared to previous algorithm.

Chapter 4 – Pressure Retarded Osmosis (PRO) is an emerging renewable energy technique with potential for scale up and commercialization. It has been extensively investigated in the last decade as a new technique for power generation from osmotic energy. Both open-loop and closed-loop PRO have been suggested for power generation, differing in that the draw and feed solutions are disposed of after membrane treatment in the open-loop PRO process. The closed-loop PRO (CLPRO) process operates like a heating engine in which the diluted draw solution enters a special unit after leaving the membrane for regeneration and reuse. CLPRO uses a pure draw solution for process operation, hence reducing membrane fouling and cleaning requirements. Furthermore, there is no waste discharge to the environment and no pretreatment is required for the feed solution.

CLPRO design requires optimizing the operating parameters to reduce the capital and operating costs of the process. Membrane area, feed pressure, draw solution, and draw solution flow rate are among the key operating parameters that significantly impact CLPRO performance. A computer model is developed in the current study to evaluate the impact of these parameters on CLPRO performance. The feed solution osmotic pressure was considered to be negligible because of the high purity of water produced from the regeneration unit. Simulation results revealed that feed pressure is the most influential parameter on the CLPRO performance, followed by the draw solution flow rate and membrane area. In fact, the membrane impact on the performance of CLPRO did not exceed 4%. The result of this study should be considered in the early stages of CLPRO design to reduce operating and capital costs.

Chapter 5 – This chapter proposes the combination of temporal processing through Rake combining, based on direct sequence-spread spectrum (DS-SS), and multiple antenna beamforming or antenna spatial diversity as a possible physical layer access technique for fifth generation (5G) small cell base stations (SBS) operating in the millimetre wave (mmWave) frequencies. Unlike earlier works in the literature aimed at previous generation wireless, the use of the beamforming is presented as operating in the radio frequency (RF) domain rather than the baseband domain, to minimise power expenditure as a more suitable method for 5G small cells. Some potential limitations associated with massive multiple input-multiple output (MIMO) for small cells are discussed relating to the likely limitation on available antennas and resultant beamwidth. Rather than relying, solely, on expensive and potentially power hungry massive MIMO (which in the case of a SBS for indoor use will be limited by a physically small form factor) the use of a limited number of antennas, complimented with Rake combining, or antenna diversity is given consideration for short distance indoor communications for both the SBS and user equipment (UE). The proposal's aim is twofold: to solve eroded path loss due to the effective antenna aperture reduction and to satisfy sensitivity to blockages and multipath dispersion for indoor, small coverage area base stations. Two candidate architectures are proposed. With higher data rates, more rigorous analysis of circuit power and its effect on energy efficiency (EE) is provided. A detailed investigation is provided into the likely design and signal processing requirements. Finally, the proposed architectures are compared to current fourth generation long term evolution (LTE) MIMO technologies for their anticipated power consumption and EE.

Chapter 6 – In virtualised 5G networks, evaluating the Power Consumption (PC) and other trade-offs could help optimising the power planning, quality of service (QoS) and Energy Efficiency (EE). In this chapter, parameterized and componentised power models (PMs) are proposed to evaluate the virtualization based software defined cloud-radio access network (VSDCRAN) paradigm. This evaluation requires a knowledge about the PC of each individual component which constitute the virtualised server. Cooling and total PC for this architecture with and without virtualisation have been compared using differentiated parameters, such as varying number of bare-metal Base Band Units (BBUs), VMs and system's bandwidth share. The results show dramatic decrease in the total PC via virtualizing the core network (CN), whereas the trade-off of such gain is extensively demonstrated.

In: Energy Efficiency
Editor: Julian E. Perez
ISBN: 978-1-53611-040-1

Chapter 1

AN ENERGY EFFICIENT ASSESSMENT TOOL FOR BUILDINGS IN DEVELOPING COUNTRIES

M. N. Addy*, PhD and E. Adinyira, PhD
Department of Building Technology,
Kwame Nkrumah University of Science and Technology,
Kumasi, Ghana

ABSTRACT

This chapter reports on the research approach adopted in the development of a building energy efficiency assessment tool for a developing country based in Sub-Saharan Africa: Ghana. The inability to solve peak demand constraints by finding a balance between reducing demand and increasing supply has resulted in power rationing coupled with erratic electricity supply in Ghana. Frequently used is the energy standard which is instrumental in enhancing efficient energy design in buildings and overall energy efficient improvements. Internationally, the field of building energy efficiency assessment methods has been characterized by a plethora of research. However, no building energy standards are in place in Ghana and this is coupled with dearth of research focused on these areas in such regions. These complexities and peculiar challenges informed the adoption of an exploratory sequential

* Corresponding Author E-mail: mljaddy@yahoo.co.uk.

(mixed method) approach. This approach consisted of a synthesis and analysis of building energy assessment methods with the Delphi Technique which was further combined with both the Delphi-Analytic Hierarchy Process and a simulation based study. Exploring various related research approaches and the undergirding research paradigms, the study discusses the use of the adopted approach against the backdrop of the region's idiosyncratic challenges. Key amongst the strength of the adopted mixed method approach is that it enabled answers to be provided to research problems that would not have been answered using just one single approach. Also, the method built a stronger case for conclusion through convergence and verification of findings and made it possible for a holistic knowledge to be provided, necessary to inform both theory and practice. This approach provides answers which were wide-ranging and broad due to the multiplicity of research questions used. A generic model for the development of a building energy efficiency assessment tool suited for Ghana and similar regions is thus provided.

1. Background

Recent economic development has propelled an increase in building energy consumption in most developing countries, African countries not being excluded. Hui (2000) averred that, developing countries are experiencing an increase in construction of buildings however, the uptake of efficient technologies are low due to the energy prices and market. In Ghana, it is projected that consumption of electricity is increasing by 10% per annum due to the surging population's demand for energy (Essah, 2011). One major challenge and critical developmental goal of most growing economies is the need to increase their energy capacity to meet their rising energy consumption base.

Ghana's population over the years has been on the increase with strides in economic growth; however, the same cannot be said of the energy situation (Essah, 2011). Statistics show that marginal increase has been seen in energy supply as compared to the burgeoning population growth (Essah, 2011). Currently, it is estimated that 55% of Ghana's capacity to generate electricity is presently attributed to hydro-based sources (Energy Commission Ghana, 2015). Electricity consumption in Ghana can be broken down into four main sectors; residential, non-residential, industrial and street lighting. As at 2015, the bulk of the energy was consumed by the industrial sector with the second highest consumer sector being the residential sector. Over the decade, the non-residential energy demand has increased by over 100% (Energy Commission

Ghana, 2015). In Ghana, electrical energy in residential and non-residential buildings is largely used for air-conditioning due to the warm humid climate. Electrical appliances and lighting also consumes the electricity used in this sector. This situation is not different from most African countries within the sub region. Over the first decade of the 21st century, energy development in Africa has not been at par with the increasing demand, placing a huge strain on the already existing resources. It is estimated that Africa's energy demand will grow by 85% between 2010 and 2040 (International Energy Outlook, 2013).

Rising energy consumption has had severe environmental impacts including increasing energy demand, global warming, air pollution and acid rain. In Ghana, power rationing coupled with erratic electricity supply has ensued due to the inability to find a balance between reducing demand and increasing supply (Ackah et al., 2014). The incessant blackouts and lack of reliable energy supply has had a negative effect on the economy resulting in job cuts and lower productivity (Doe and Asamoah, 2014). At the local level in many African nations, peak demand is often met with power from privately owned diesel generators, which supply very expensive electricity and pollute the air. In most developed countries, there is the usage of building energy standards to curb some of these impacts. Energy regulation is one of the most frequently used instruments for energy efficiency improvements in buildings (OECD, 2013). However, Iwaro and Mwasha (2010) unearthed in their study that more than 40% of emerging markets do not have a building energy standard in place, 20% have mandatory, 22% have mixed and 16% proposed. As identified by Iwaro and Mwasha in the year 2010, Ghana is amongst developing countries lacking a building energy assessment tool (Iwaro and Mwasha, 2010). This lack of an energy assessment tool necessitated a study into the development of a tool to be used in Ghana.

Consequently, a study has been conducted and resulted in the development of a building energy efficiency assessment tool (BEEAT) to be used in predicting the energy efficiency of office buildings in Ghana (Addy, 2016). This chapter focuses on the research approach used in the development of the tool. The research was focused in Ghana but is applicable to similar developing countries grappling with building energy problems. The study into the development of the tool saw the carving out of four main objectives. These objectives provide a series of step that led towards the development of the final tool. In order to achieve each step identified a specific research method was adopted. Already Fellows and Liu (2008) have proffered that the selection of a suitable research method is a major determinant in the success or otherwise of any research activity. Firstly, an overview of research paradigms and strongly related research approaches are presented. This is followed by the research design and methodology adopted. Thereafter, a discussion of the chosen research approach against the peculiar challenges within the region and how it results in a bespoke tool is presented.

2. Research Paradigm and Approach

Yin (2009) encapsulates research design as the logical arrangement that links the collated empirical data to the initial objectives of a research and consequently to its conclusions. The multifaceted and multi-disciplinary nature of building energy assessment studies reflects its un-integrated theory. Thus, building energy assessment often draws on a wide range of theoretical basis from the social and natural sciences. Researchers are met with a difficult problem in clearly articulating an appropriate research design due to this multiplicity of diverse theoretical basis.

Many considerations come to play in the decision to select a suitable research methodology for a study. Most influential among those considerations is the defining aim of the research study and the data type that should be collected. The researcher is confronted with two broad research methodologies which must be deliberated upon in the subsequent selection of the appropriate one for the study. The research methodologies can be categorized as quantitative and qualitative. However, a use of either one or a combination of both is not uncommon in literature as far as the adopted approach is adequate in answering the research questions.

Research methodology is not developed in a vacuum. Various paradigms guide and inform a given research in the selection of the appropriate research methodology (Liu, 2008). Paradigms provide insight into the way researchers seek to comprehend and make sense of reality (Maguire, 1987). Paradigm shapes the way researchers perceive the research methodology adopted and the techniques to be used. Two main research paradigms dominate literature: positivism and interpretivism (Oates, 2006). In positivism, social phenomenon is assumed to follow natural laws and thus quantitative logic can be used. The scientific method employed avoids the inclusion of human influence in establishing the meaning of reality and as such, prevents researchers' bias (Guba, 1990). On the other hand, interpretivsim assumes that social phenomenon does not follow natural laws but it is constructed based on peoples' interaction with their environment (Walliman, 2001). This thus assumes the use of qualitative logic. The use of these paradigms is influenced by the ontological, epistemological and axiological assumptions underpinning the research (Keraminiyage et al., 2005). Consequently, the decision on the epistemology to use is greatly influenced by the ontological consideration and guided by the axiological positions of the research question.

To be able to develop the tool, an exploratory mixed methodology approach was adopted and this was due to the natural fluidity of the field of sustainable assessment methods as it includes a wide range of criteria from different ecological and economical dimensions (Ding, 2008). Further, the adoption of a mixed method Approach enabled answers to be provided to research problems that would not have been answered using just one single approach. The method built a stronger case for conclusion through convergence and verification of findings. It also made it possible for a holistic knowledge to be provided and this is necessary to inform both theory and practice. This approach provides answers which were of wide range and broad, due to the multiplicity of research questions used. The study adopted a synthesis and analysis of building energy assessment methods with the Delphi Technique which was combined with both the Delphi-Analytic Hierarchy Process and a simulation based study in the current research.

3. Research Design

The next step after selecting a suitable method based on the necessary philosophical paradigm is to decide on the research design. The selected research design will consequently influence the selection of research instruments to be employed (Sarantakos, 2005). Guided by the aim and the research objectives, the decision on the research design should be made. The research objectives in question explore a sustainable and ecological context commonly regarded as a multi-dimensional issue (Ding, 2008). Based on the above considerations, the study adopted the mixed methods research (quantitative and qualitative combined) approach as previously stated. This method provides a robust procedure of meeting the research aim and objectives. Subsequently, to meet the stated aim of the study as outlined at the outset of the research, the following strategies were adopted:

3.1. Stage One: Comparison of Well-Known Methods

Previous researches show that the comparison method has been used for the development of tools in various countries (Poston et al., 2010; Ali and Al-Nsairat, 2009; and Zuhairuse et al., 2009). Most environmental assessment methods have all been designed to fit a specific region and certain environmental factors might hinder the direct usage of any current environmental assessment tool or method (Cooper, 1999; Crawley et al., 1999). Cole (2005) opined that it is imperative that the initial stage of a new environmental tool development is a comparative analysis of established methods. Hence, the first stage of BEEAT development was a comparative analysis of the most significant and internationally recognized environmental assessment methods (Alyami and Rezgui, 2012). Different tools are compared to identify gaps, parallels and differences that may exist between them. The comparisons orientation is skewed towards the development of the energy assessment tool for Ghana.

The objective for which differing tools are chosen is to investigate the impact on the countries from which they originated from. The presentation of every tool constitutes an aspect of a world leading to diversity in environmental, economic, social and cultural aspects. The selection of these methods relied on the credibility of the organizations which launched and operated them and their success in the marketplace. Two main primary sources of information were used for the comparative analysis: related publications that analyze and compare their components and technical manuals of the selected methods. A number of areas were assessed and these included the scope of assessment, assessment method and the various ways in which calculation was computed for the tools.

The multiplicity of building assessment methods poses a difficult challenge in comparing different tools due to various aspects that should be taken into consideration (Poston et al., 2010; Papadopoulos and Giama, 2009; Haapio and Viitaniemi, 2008). Thus, to give the specific comparison of the designated assessment methods, the research categorizes the criteria of every rating tool into three classifications based on the three dimensions of sustainability: environmental, economic and social. This enabled easy comparison and afforded the advantage of being able to build a case for adopting, adapting or rejecting a particular aspect of a tool.

3.2. Stage Two: Delphi Technique

The Delphi technique was developed in the early 1950's principally by Dalkey and Helmer (Miller, 2006) and is structured in a way to solicit the opinion of experts on complex issues by structuring the communication amongst them (Linstone and Turoff, 1975). The structure in this technique helps to achieve consensus and brings about stability in the group judgment even with subjective issues. The technique in the Delphi is founded on four fundamental principles (Rowe and Wright, 1999; Alder and Ziglio, 1996; Linstone and Turoff, 1975): Iteration, Anonymity, Controlled feedback, Statistical group responses.

The focus of the study is towards the development of a building energy assessment tool. This focus involves different topics and necessitates the use of more than one research method. The initial step taken in the study has been to review existing methods with the aim of finding a basis to develop a novel tool. The next research method was the adoption of a consensus based approach to create an applicable building assessment criteria. This approach was motivated by the multi-dimensional nature of the research topic, exploring both sustainable and ecological contextual areas (Ding, 2008). Chew and Das (2008) posit that the drive for scientific evidence within such a subject requires the use of a consensus-based approach in arriving at a more appropriate building energy assessment method. This consequently led to the selection of the Delphi method.

The Ranking Delphi technique was utilized in arriving at the various factors that influence energy efficiency standards in buildings. Organized in two rounds, key actors in the field of building energy were consulted in Ghana to arrive at an appropriate energy efficiency method. Following in the steps of a successful Delphi, procedures have been adhered to ensure panel suitability in relation to both size and composition (Okoli and Pawlowski, 2004; Dalkey and Helmer, 1963). The Delphi survey used in this research looked at a target number of 30 members for the panel. The panel comprised professionals from academia, government and industry. This process culminated in the identification of criteria and categories to be used in the development of the final BEEA Tool.

3.3. Stage Three: Delphi-Analytic Hierarchy Process

Though the Delphi method is a useful instrument for brainstorming and evaluating crucial factors, results from the Delphi may require additional research in managing moderate consensus among professionals. The AHP which was initially instituted by Saaty, can be the supplementary research for a Delphi study (Filtvedt et al., 2012). The AHP is an extensively used instrument in multi-criteria decision-making. Additionally, the Delphi consultation may be implemented prior to the AHP to refine a set of factors. This increases the validity of the AHP as posited by Joshi and Shankar (2011). They state that in an AHP-compared set, the number of factors should be plus seven or minus two. It is very usual to use AHP to choose technologies in modern scientific research. Studies abound in this direction. For example, a study by Joshi and Shankar (2011) focused on developing a based benchmarking framework using the AHP. The framework was for the purposes of evaluating the performance of a firm and further prioritize alternatives within the said firm. Another study by Kim et al., (2013) used the Delphi-AHP methods in choosing the primacies of a waste electrical and electrical equipment within the field of waste management. Further Daw et al., (2013), combined the Delphi and fuzzy AHP to select technology of late-starters in the energy-smart photovoltaic industry. The wider framework of this selection was to serve as a guide in industrial technologies procurement and resource allocation. The AHP method was made use of in the selection of the most appropriate alternative from amongst numerous efficacious solutions.

In this research approach following the already selected Delphi technique, another Delphi-AHP based methodology is adopted to create a qualitative and quantitative measurement system, when choosing and prioritizing vital indicators for predicting building energy efficiency. To achieve this, the first round consists of question posed to identify the indicators of building energy efficiency. In the second round of the Delphi, feedback of the first round is given to the respondents and subsequently questions posed using Saaty's scale are provided. This enabled the comparison of the indicators using the analytic hierarchy process. The Delphi-AHP is utilized to develop a suitable weighting system, for the approved categories that resulted from Delphi method. Questions posed were based on proposed criteria with options provided to investigate participant's agreement or disagreement. The degree of consensus reached among the experts was measured using statistics such as mean, median and interquartile range. The results from the first round fed into the round two. The output of this process was the development of a proposed grading system to be used by the tool.

3.4. Stage Four: Simulation Based Study

To achieve the last objective a simulation exercise based on a case study was conducted. Dynamic thermal modelling was used to simulate a generic office building in the hub of Accra, Ghana. The office building was then again simulated based on best practice to serve as a reference building for the bases of comparison. A single case study was used and involved the analysis of a high-rise office based on the various identified categories and criteria from the Delphi and DHP survey. The specific aim being to make practicable the proposed grading system, exploring both its strengths and weakness. The approach used was the initial analysis of the building design and construction fabric and materials. As-built drawings including plans were obtained for the identified office. In addition, annual electricity bills for the selected building were also obtained from the facility's management office of the building. This was followed up with an interview of occupants of the offices to understand how the building is used. Many reasons informed the selection of the office building. Firstly, the selected building needed to be a generic building reflecting current trends in office buildings with respect to high rise office buildings in Ghana. Consequently, the selected building should mirror typical office building in terms of occupant density, size and area. Another requirement was the ability to obtain data with ease. This includes architectural drawings, construction details, occupant usage and if possible, electricity bills. Based on the above considerations, the selected case study was a high-rise office building located in the city center of Accra (the capital city of Ghana), known as the Ridge Towers (RT).

Three main sets of data were collated and this included the climatic data, information on the building parameter and how the building operates. Information on the building characteristics and parameters were collected between the period of about one year starting in May of 2012 and ended in April 2013. During this time the Accra Meteorological Department was successfully consulted for outdoor weather data. Due to lack of comprehensive data on outdoor weather information, segments of a synthetic weather file for Accra was identified and used. This approach has been used in previous studies within the Ghanaian context (see for example Amos-Abanyie, 2012 and Koranteng et al., 2009). From this, it was possible to simulate hourly, daily, monthly and annual energy consumption figures. To simulate and examine the energy consumption of the case study building, the DesignBuilder tool was employed. The tool provides advanced modelling tools in an easy-to-use interface.

4. Providing a Holistic Approach to Sustainable Development in Designing a Building Energy Efficiency Tool

One key deliverable of an assessment method is improved environmental performance. With increase in understanding of the concept of sustainability globally, future needs point to a more comprehensive sustainable method (Haapio and Viitaniemi, 2008; Sebake, 2009). Currently, most assessment tools place more emphasis on environmental issues like energy landscape, resources, emissions and the indoor environment quality (Sinou and Kyvelou, 2006; Poston et al., 2010). Undeniably, environmental issues are a global concern and attention ought to be paid to it. This goal most current tools perform excellently on (Mao et al., 2009; Poston et al., 2010). Nevertheless, other aspects of sustainability should not be neglected to the detriment of delivering a holistic sustainable method (Ding, 2008; Haapio and Viitaniemi, 2008).

Extant studies show key parameters in achieving sustainable performance include consideration of economic, social and cultural aspects (Forsberg and von Malmborg, 2004; Sinou and Kyvelou, 2006; Haapio and Viitaniemi, 2008; Mao et al., 2009). Todd et al., (2001) in a study provide that these parameters may present not only impediments but also opportunities for developing a rating system among developing countries. Moreover, Poston et al., (2010) cited that criticism is also levelled against assessment tools for focusing on environmental criteria at the cost of social and economic criteria despite a shift towards more sustainable assessment. It is held that the local context is a major determinant of the significance of economic, social and cultural aspects. This is largely due to its variation across countries. Case in point Alyami (2015) argues that social and cultural aspects play a pivotal role in Arabic countries such as Saudi Arabia. This is supported by Cole (2005) who posits that for developing countries, social and economic aspects rank high. Arguably environmental concern is largely at the forefront in advanced countries. Hence, it is realized that Reed et al., (2006) regard economic, social and cultural dimensions as exclusive features which could stop a tool take-up. Significance of this sort could be inferred which consists of differing studies (Ali and Al Nsairat, 2009; Zuhairuse et al., 2009; Al-Sallal et al., 2013) that have created environmental evaluation tools for various developing countries. It is perceptible that some developing countries have commenced the development of domestic assessment methods with suitability for their milieu, and economic, social, cultural and historical circumstances (Todd et al., 2001).

Evidently, various studies have debated the inclusion or otherwise of economic, social and cultural aspects of a building assessment method. Moving forward, these issues have a great potential of critically affecting existing environmental assessment methods such as Building Research Establishment Environmental Assessment Method (BREEAM) and Leadership in Energy and Environmental Design (LEED). For instance, the larger number of countries commence their assessment method instead, with the usage of accessible existing tools resulting from their lack for cultural aspects. Proof has been provided by a new generation of assessment methods springing up because of upsurges in the demand to consider economic and cultural dimensions in assessment tools such as Estidama (Poston et al., 2010).

5. A Case for the Adopted Research Methodology

A critical element in the development of building energy tools is the existence of energy codes and data studies. Studies corroborate this element. For example, in the development of energy label in Singapore, the basis for judgement was the code already existing (Eang and Priyadarsini, 2008). Countries that have adopted these tools have frameworks utilized in developing the tool (Bagheri et al., 2013; Batista et al., 2011). Also, the existence of building energy standards define the framework for more studies and drive research into many other areas. Janda (2008) in their study show that most African countries lack building energy standard. A study in Iran exploring energy label for office buildings was based on their existing standards (Bagheri et al., 2013). In Ghana, firstly, there are no existing energy efficiency codes for buildings. Secondly, lack of energy data poses a difficult challenge. Due to these impediments, the energy efficiency categories and criteria will serve as a starting point in the prediction of energy efficiency. The use of standard practice is also adopted to serve as benchmarks in the use of the tool.

Energy efficiency is the driving force of implementing environmental assessment methods. This is because the building industry worldwide is putting high pressure on natural resources, embodied in the tremendous demand for electricity. As mentioned earlier, electricity demand is increasing at a record level; and it is expected that the next 30 years will witness higher rates worldwide (World Energy Outlook, 2006). However, building environmental assessment methods are still evolving as a field of investigation. The existing assessment methods are heavily criticised for the absence of integrating regional variations (Haapio and Viitaniemi, 2008). Thus, it is of vital importance for policy makers and other building professionals (assessors) to detect how much environmental and economic benefits can be gained from the implementation of sustainability measures. It is exactly this degree of specificity that this study attempts to add to assessment methods by focusing on regional climate conditions. It can be argued that environmental assessment methods designed in the manner described above can bring about a new appreciation of climate influence on the built environment. They can provide decision makers with a robust strategy aiming to classify regional climates and in turn direct the growth and potential building boom to these areas. Again, the approach used does not eliminate the wider socio-economic environment but provides the recognition of variables necessary for the effective use of the tool.

The availability of data is a major determining factor for the success of energy research. This is mainly due to the reason that the research relies heavily on data including building energy consumption profiles, general building characteristics, weather data and the like (Lee and Kung, 2011). Bagheri et al., (2013) in their study posited that similar studies within the direction of standard and labelling procedures for buildings worldwide mostly considered a sample society of buildings as the data bank for studies. It has been noted that lack of data for building energy consumption remains a significant drawback for extracting general conclusions on the energy performance of the building stock (Nikolaou et al., 2009). Lack of energy data and dearth of studies within the research direction remains a critical challenge in Ghana. The peculiar nature of developing countries, particularly within the African context - increasing urbanization and the pervasiveness of poverty - presents a unique mix to the situation. This is exacerbated by the lack of energy codes and standards, lack of energy data and studies. Thus, in finding solutions to the problems, the research approach adopted needs to be able to resolve these issues.

6. Finding Answers Resolving Difficulties

Following the discussion, three major issues need to be deliberated upon in choosing the appropriate research method to meet the research aim for the direction of the study. Firstly, the research method must be able to provide a viable starting point and consequently identify key variables associated with energy assessment tool. Next, the research method should be able to ascertain the correlation between the variables identified and how energy efficiency is predicted by these variables. Lastly, they must allow for detailed data to be collated and analyzed, showing how the identified variables can be used in determining energy efficiency.

The two main research methods, quantitative and qualitative, both have inherent weaknesses and strengths. Critics of quantitative methods posit that such methods lack contextual realism whereas qualitative methods have a weakness of not being able to study with a level of accuracy and is criticized for inefficiency in providing accurate study of variable relationship (Sarantakos, 2005; Tashakkori and Teddlie 2003). In developing an energy assessment tool for Ghana, the use of only one approach would be limiting. This requires the use of an approach that combines the strengths of both qualitative and quantitative methods, i.e., a mixed method approach. This approach provides the researcher with the advantage of discovering and justifying key components within a study. For example, the use of qualitative research provides the necessary realism and depth of information for hypothesis formulation and theory building due to its involvement of people (Tashakkori and Teddle, 2003). Qualitative study gives the researcher the information depth provided in detailed data, rich for his interest. By using techniques such as the Delphi survey and data review, the contextual nature of the research question is captured. In addition to the above, as sustainable building assessment criteria are normally thought to be of a multi-dimensional criteria (Ding, 2008), scientific proof indicates that a consensus-based approach is most suitable for the development of all-inclusive and efficacious building environmental assessment categories and criteria (Chew and Das, 2008). Building energy data studies such as a simulation based study also provide the required framework to properly develop the tool.

Consequently, it should be noted that this is a first step in developing the tool. Studies show that closing the knowledge gap in building energy efficiency studies in the Ghanaian environment will ultimately call for revisions to be made. As such, further revisions are expected based on empirical findings from studies of the proposed tool. In addition to the above, the development of the tool was based on the building energy experts in the field. This has its drawbacks and it is expected that key stakeholders will be involved when revisions are necessitated. The lack of standards necessitated the use of established benchmarks. Due to the difficulty in getting data, some variables such as aeration and indoor air quality assessment and air change rates have been apportioned standard and approximated values. Future development on the methods of rating ought to consider circumstances and matters associated with indoor setting to guarantee that energy efficacy does not ever complicate the quality of the indoor spaces. This is a first step in the right direction.

7. LOOKING AHEAD

Energy efficiency improvement remains critical in solving the energy problem. Moving forward, it is imperative to explore how the extant challenges can be solved within the study region. The discussions follow on.

7.1. The Need for Expertise and Requisite Knowledge

A strong knowledge base in building energy tools and building energy simulation is crucial in the direction of this research. Bagheri et al., (2013) averred that the studies are done either by the application of the energy simulator software tools or by using the energy audit instruments. In addition, in many developed countries, software tools are used to estimate the energy performance grade for an existent or designed (not built) building. Building the local capacity with the use of building energy simulation skills and other soft skills required will enhance the ability to further carry out research of this nature. Secondly, the availability and accessibility of building energy experts is key as the process of development of a tool or its assessment has the target population to be building energy experts. This assertion is supported by Bagheri et al., (2013). In their work, they targeted energy experts to do the evaluation. Many research methodologies chosen, unquestionably, can be performed on only experts such as Delphi survey and analytical hierarcy process. Also, energy audit can only be performed by trained personnel.

The capability of a country to introduce and sustain energy efficient buildings is the result of a successful knowledge base in this arena. Yik and Lee (2002) in their paper reiterate that the lack of knowledge and skills to identify causes of inefficiency and to devise mitigation measures are key knowledge barriers to building energy efficiency. Qian et al., (2006) in China also noted that many Chinese rarely invest in energy-efficient products due to little knowledge about the benefits and co-benefits of energy- efficient buildings. In developed countries, there is a robust energy-saving technical evaluation system; which feeds into the design of energy efficient buildings. It is critical to gather and keep information on best practices in order not only to learn lessons, but foster a growth in the area of building energy efficiency. It is critical that a building designer has the expertise to design energy efficient buildings

7.2. The Role of Policy Instruments

The Energy Performance Building Directive (Directive 2002/91/EC) introduced the compulsory energy certification of buildings in the European Union from 2006 and it has played a significant role in the common policy to monitor and reduce energy consumption. The introduction of this directive has also directed more research into energy buildings as member states of the European commission explore ways of meeting this directive. The legal and regulatory environment is the uppermost factor that affects the implementation of building energy efficiency. Poor participation of government and inadequate policies cripple the move for energy efficient buildings. Studies has shown that policy plays a vital role in ensuring building energy efficiency is achieved.

The need for government involvement in the adoption of energy efficient buildings is imperative. This is because, enforcing energy efficiency through regulatory control has the advantage that all parties concerned will be aware of the requirements, and ensures that a certain minimum level of performance is achieved. The development of a building energy standard and its accompanying policy and regulatory framework has been the spring board of the development of energy efficient measures in many developed countries. The presence of this will drive this agenda towards a positive direction.

7.3. Providing Access to Finance

The experience of some European countries shows that special funds play a crucial role in promoting energy-saving work. The lack of a funding source for financing the required improvement measures is obviously a barrier, which occurs when insufficient reserve has been made by the building management, or no loans are available from external sources (Yik and Lee, 2002; Painuly et al., 2003). Qian et al., (2006) posited that the lack of capital to buy energy-efficient equipment and the high cost of alternative technologies are all barriers that impede the take up of energy efficient buildings. To promote the design of energy efficient buildings in our context, access to finance must be improved.

Conclusion

This chapter encompassed the presentation of the methodology made use of in the conduct of a study. On the basis of the study's research agenda, an exploratory mixed methodology approach was adopted. It was identified that the adoption of the mixed method approach enabled answers to be provided to research problems that would not have been answered using just a single approach. Similar studies across the globe have adopted the use of group consensus measurements such as the Delphi and AHP, while others have also adopted both the Delphi and AHP (Alaymi, 2015; Bagheri et al., 2013; Beradi, 2012). However, the current study due to peculiar issues within the region adopted the use of both the Delphi Method with the Delphi-AHP. The developed tool was later subjected to a simulation study. This unique methodology - the combination of the Delphi survey, the Delphic Hierarchy Process and the simulation study - differentiates it from other similar studies conducted. Further, the approach afforded a bespoke tool fit for the peculiar region to be developed. Through this study, a robust methodology tailored for regions characterized by urbanization with a growing economy combined with lack of energy data studies have been developed and used.

The need for stimulating the market demand for sustainable practices in the built environment requires the large-scale adoption of adapted environmental building assessment methods. While western countries have widely engaged in this avenue, developing economies such as Ghana are still trailing behind. Also, Ghana presents a great potential for renewable energy use. Hence, it becomes imperative to design and put into operation a benchmark scheme which will assess the principles of sustainable building energy consumption. This will foster the recognition of green building principles with the extensive adoption of sustainable energy practices. In providing solutions to Africa's sustainable development, energy efficiency forms a major integral part.

REFERENCES

Ackah, I., Adu, F. and Takyi, R. O. (2014). On the Demand Dynamics of Electricity in Ghana: Do Exogenous Non-Economic Variables Count? *International Journal of Energy Economics and Policy* 4(2), 149-153.

Addy, M. N. (2016). Development of a Building Energy Efficiency Assessment Tool for office buildings in Ghana. Ph.D Kumasi: Kwame Nkrumah University of Science and Technology.

Adler, M. and Ziglio, E. (1996). Gazing into the Oracle: The Delphi method and its application to social policy and public health. London: Kingsley Publishers.

Alessandro, M, Rosano, M. B. and Simeoni, P. (2013). A decision support system for sustainable energy supply combining multi-objective and multi-attribute analysis: An Australian case study. *Decision Support Systems* 57, 150–159.

Ali, H. H. and Al Nsairat, S. F. (2009). Developing a green building assessment tool for developing countries – Case of Jordan. *Building and Environment,* 44(5), 1053-1064.

Al-Sallal, K, Al-Rais, L and Dalmouk, M. B. (2013). Designing a sustainable house in the desert of abu dhabi. *Renewable Energy* 49, 80-84.

Alyami, S. (2015). The development of Sustainable Assessment Method for Saudi Arabia Built Environment. Ph. D. Cardiff University.

Alyami S. H., Rezgui, Y., and Kwan, A. (2014). The development of sustainable assessment method for Saudi Arabia built environment: weighting system. *Sustainability Science* 10(1), 167-178.

Alyami, S. H., and Rezgui, Y. (2012). Sustainable building assessment tool development approach. *Sustainable Cities and Society* 5, 52–62.

Amer, M. and Daim, T. U. (2011). Selection of renewable energy technologies for a developing county: a case of Pakistan. *Energy Sustainable Development* 15(4), 20-35.

Amos-Abanyie, S. (2012). Effects of thermal mass, window size and night-time ventilation on peak indoor temperature in the warm humid climate of Kumasi, Ghana. Ph. D. Kumasi: Kwame Nkrumah University of Science and Technology.

Andaloro, A. P. F., Salomone, R., Ioppolo, G. and Andalor, L. (2010). Energy certification of buildings: A comparative analysis of progress towards implementation in European countries. *Energy Policy* 38, 5840-5866.

Bagheri, F., Mokarizadeh, V. and Jabbar, M. (2013). Developing energy performance label for office buildings in Iran. *Energy and Buildings* 61, 116-124.

Batista, N. d. N., Rovere, E. L. and Aguiar, J. C. R. (2011). Energy efficiency labeling of buildings: An assessment of the Brazilian Case. *Energy and Buildings* 43, 1179-1188.

Berardi, U. (2012). Sustainability assessment in the construction sector: rating systems and rated buildings. *Sustainable Development* 20, 411-424.

Chang, K. F., Chiang, C. M., and Chou, P. C. (2007). Adapting aspects of GBTool 2005—Searching for suitability in Taiwan. *Building and Environment* 42, 310–316.

Chew, M. Y. L. and Das, S. (2008). Building grading systems: A review of the state-of the art. *Architectural Science Review* 51(1), 3–13.

Cobbinah P. B., Erdiaw-Kwasie, M. and Amoateng, P. (2015). Rethinking sustainable development within the framework of poverty and urbanisation in developing countries. *Environmental Development* 13, 18-32.

Cole, R. (2005). Building environmental assessment methods: Redefining intentions and roles. *Building Research and Information* 33(5), 455–467.

Cooper, I. (1999). Which focus for building assessment methods - Environmental performance or sustainability? *Building Research and Information*, 27(4-5), 300-308.

Crawley, D. and Aho, I. (1999). Building environmental assessment methods: Applications and development trends. *Building Research and Information*, 27(4-5), 300-308.

Dalkey, N. C. and Helmer, O. (1963). An experimental application of the Delphi method to the use of experts. *Management Science* 9 (3), 458-467.

Daw, M., Chia-Chin C. and Shiu-Wan H. (2013). The selection of technology for late-starters: A case study of the energy-smart photovoltaic industry. *Economic Modelling* 35, 10-20.

Ding, G. K. C. (2008). Sustainable construction: The role of environmental assessment tools. *Environment Management* 86, 451-64.

Doe, F. and Asamoah E. S. (2014). The effect of electric power fluctuations on the profitability and competitiveness of SMEs: A study of SMEs within the Accra Business District of Ghana. *Journal of Competitiveness* 6 (3), 32-48.

Eang, L. S. and Priyadarsini, R. (2008). Building energy efficiency labelling programme in Singapore. *Energy Policy* 36, 3982-3992.

Energy Commission Ghana (2015). National Energy Statistics – 2015. Accra: Energy Commission Ghana.

Energy Commission Ghana (2007). Strategic National Energy Plan. Accra: Energy Commission Ghana.

Energy Consumption Ghana (2007). Ghana Energy Outlook. Accra: Energy Commission Ghana.

Fellows, R. F. and Liu, A. M. M. (2008). Research Methods for Construction. 3rd ed. Oxford: Blackwell Publishing Limited.

Filtvedt, W. O., Holt, A., Ramachandran, P. A. and Melaaen, M. C. (2012). Chemical vapor deposition of silicon from silane: review of growth mechanisms and modeling/scaleup of fluidized bed reactors. *Sol Energy Mater Sol Cells* 107, 188-200.

Forsberg, A. and von Malmborg, F. (2004). Tools for environmental assessment of the built environment. *Building and Environment* 39(2), 223-228.

Guba, E. G. (1990). The paradigm dialog. California: Sage Publications.

Haapio, A. and Viitaniemi, P. (2008). A critical review of building environmental assessment tools. *Environmental Impact Assessment Review* 28(7), 469–482.

Hui, S. C. M. (2000). Building energy efficiency standards in Hong Kong and mainland China. In: Proceedings of the 2000 ACEEE Summer Study on Energy Efficiency in Buildings, 20–25 August 2000, Pacific Grove, California.

Janda, K. B. (2008). Worldwide status of energy standards for buildings. In: Proceedings of the Fifth Annual IEECB, April 9-10, 1761-1769. Frankfurt Germany.

Joshi, R., Banwet, D. K. and Shankar, R. (2011). A Delphi-AHP-TOPSIS based benchmarking framework for performance improvement of a cold chain. *Expert Systems with Applications* 38(8). 10170-10182.

Iwaro J, and Mwasha A. (2010). A review of building energy regulation and policy for energy conservation in developing countries. *Energy Policy*. 2010; 38: 7744–7755.

Keraminiyage, K. P., Amaratunga, R. D. G and Haigh, R. P. (2005). A capability maturity approach for construction projects improvement: use of case studies approach. In: 5th Postgraduate Conclusions and Recommendations Research Conference, University of Salford, April 13th -15th, 465-475.

Kim, M., Jang, Y. and Lee, S. (2013). Application of Delphi-AHP methods to select the priorities of WEEE for recycling in a waste management decision-making tool. *Environmental Management* 128, 941-948.

Koranteng C., Mahdavi A, Orehung, K. and Proglhof, C. (2009). Thermal performance of office buildings in Ghana. In: Proceedings of the 11th International IBPSA Conference. Glasgow, Scotland July 27-30, 2009.

Lee, W. and Kung, C. (2011) Using climate classification to evaluate building energy performance. *Energy* 36, 1797-1801.

Lee, W. L., Yik, F.W.H. and Burnett, J. (2003) A method to assess the energy performance of existing commercial complexes. *Indoor and Built Environment* 12 (5), 311–327.

Liang, J., Li, B., Wu, Y., Yao, R. (2007). An Investigation of the Existing Situation and Trends in Building Energy Efficiency management China. *Energy and Buildings* 39 (10), 1098-1106.

Linstone, H. A. and Turoff, M. (1975). Introduction. In: H. A. Linstone, & M. Turoff (Eds.). *The Delphi method: Techniques and applications.* Reading, MA: Addison-Wesley Publishing Company.

Liu, R. F. A. A. (2008). Research Method for Construction. Singapore: Wiley-Blackwell.

Maguire, M. (1997). Doing participatory research: a feminist approach. Amherst, MA: The Center of International Education, University of Massachussetts.

Mao, X., Lu, H. and Li, Q. (2009). A comparison study of mainstream sustainable/green building rating tools in the world. In: *Management and Service Science,* 2009. MASS '09. International Conference on, 20-22 Sept. 2009, 1-5.

Miller, L. E. (2006). Determining what could/should be: The Delphi technique and its application. Paper presented at the meeting of the 2006 annual meeting of the Mid-Western Educational Research Association, Columbus, Ohio.

Nikolaou, T., Skias, I., Kolokotsa, D. and Stavrakakis, G. (2009). Virtual building dataset for energy and indoor thermal comfort benchmarking of office buildings in Greece. *Energy and Buildings* 41, 1409-1416.

Oates, B. J. (2006). Researching Information Systems and Computing. London: *Sage Publications*.

OECD (2003). Cool appliances: Policy strategies for energy efficient homes. Paris Cedex: OECD/IEA. 2003.

Okoli, C. and Pawlowski, S. D. (2004). The Delphi method as a research tool: an example, design considerations and applications. *Information & Amp. Management* 42, 15-29.

Papadopoulos, A and Giama, E. (2009). Rating systems for counting buildings' environmental performance. *International Journal of Sustainable Energy* 28(1-3), 29-43.

Poston, A, Emmanuel, R and Thomson, C (2010). Developing holistic frameworks for the next generation of sustainability assessment methods for the built environment. In: Egbu, C. (Ed) Procs 26th Annual ARCOM Conference, Leeds, UK: Association of Researchers in Construction Management, 1487-1496.

Qian, K., Wu, J. And Chan, E. H. (2006). Policy Deficiencies in Promoting Building Energy Efficiency in Mainland China. In The CRIOCM 2006 International Symposium on Advancement of Construction Management and Real Estate. Beijing.

Reed, M., Fraser, E. and Dougill, A. (2006). An adaptive learning process for developing and applying sustainability indicators with local communities. *Ecological Economics* 59, 406–418.

Reilly, D., Duffy, A., Willis, D. and Conlon, M. (2013). Development and Implementation of a Simplified Residential Energy Asset Rating Model. *Energy and Buildings* 65, 159- 166.

Rowe, G. and Wright, G. (1999). The Delphi technique as a forecasting tool: issues and analysis. *International Journal of Forecasting* 15, 353-375.

Sarantakos, S. (2005). Social research. 3rd ed. Basingstoke: Palgrave Macmillan. Satsangi.

Sebake, T. (2009). An overview of green building rating tools. Green building handbook South Africa, 1 (A guide to ecological design), 27-34.

Simons, B., Koranteng, C. and Ayarkwa, J. (2015). Simulation-based assessment of the thermal performance of high rise office buildings in Ghana. In Badu, E., Ayarkwa, J., Ahadzie, D., Adinyira, E., Owusu-Manu, D. and Kwofie, T. (eds) Procs 4th International Conference on Infrastructure Development in Africa, 25-26 March 2015, Kumasi, Ghana, pp. 44-66.

Sinou, M. and Kyvelou, S. (2006). Present and future of building performance assessment tools. Management of Environmental Quality: *An International Journal*, 17 (5), 25-41.

Tashakkori, A. and Teddlie, C. (2003). *Handbook of Mixed Methods in Social & Behavioral Research.* Thousand Oaks: Sage Publication.

Todd, J. A. (2001). Comparative assessment of environmental performance tools and the role of the Green Building Challenge. *Building Research and Information* 29(5), 324–335.

Walliman, N. (2001). *Your Research Project: a step-by –step guide for the first time researcher.* London: Sage Publications.

Wang, S, Yan, C. and Xiao, F. (2012). Quantitative energy performance assessment methods for existing buildings. *Energy and Buildings* 55, 873-888.

World Energy Outlook (2006). Measuring progress towards Energy for All. *International Energy Agency*, Paris; 2006.

Yang, Y., Li, B. and Yao, R. (2010). A method of identifying and weighting indicators of energy efficiency assessment in Chinese residential buildings. *Energy Policy* 38, 7687–7697.

Yik, F. W. H. and Lee, W. L. (2002). A preliminary inquiry into why buildings. Transactions, The Hong Kong Institution of Engineers, HK, 9(1):32–6.

Yin, R. K. (2009). Case study research design and methods; Applied social research methods series. 4th ed. London: SAGE Publications.

Zuhairuse, M. D, Nor, A. H., Elias, S., Lim, C. H., Abdul Khalim, A. and Siti, N. A. (2009). Development of rating system for sustainable building in Malaysia. In: *WSEAS Transactions on Environment and development*, March, Vol. 5, 260-272.

In: Energy Efficiency
Editor: Julian E. Perez
ISBN: 978-1-53611-040-1

Chapter 2

ENERGY RECOVERY FROM WASTE AS AN INTEGRATED MUNICIPAL SOLID WASTE MANAGEMENT: A REVIEW OF WASTE-TO-ENERGY TECHNOLOGIES

Diego Moya[1,2,*], Andrés Hidalgo[2], Juan Paredes[2] and Prasad Kaparaju[1]

[1]Griffith School of Engineering, Griffith University, Queensland, Australia
[2]Área de Energía, Facultad de Ingeniería Civil y Mecánica, Universidad Técnica de Ambato, Ambato, Ecuador

ABSTRACT

This chapter/research review is a study of the different waste-to-energy technologies (WTE-Ts) developed to date. This study of the technologies is divided into four groups: biological treatment of waste; thermal treatment of waste; incineration of waste; and landfill gas utilisation. Furthermore, integrated solid waste management systems (ISWM-S) with WTE-T are studied and some worldwide examples are provided.

Keywords: integrated solid waste management systems, municipal solid waste, solid waste management (SWM), waste-to-energy technologies

[*] Corresponding author: diego.moyapinta@griffithuni.edu.au, da.moya@uta.edu.ec (D Moya).

INTRODUCTION

A report from The World Bank estimates that currently 1.3 billion tonnes of waste is generated per year all over the world; and by 2025 this amount will increase to 2.2 billion tonnes per year [1]. These data show an urgent need for treatment of municipal solid waste (MSW) around the globe. However, although in developed countries waste is used by a resource to produce energy, heat, fuel and compost, in developing countries the collection, transport and disposal of waste are current issues [2-7]. This chapter identifies ISWM-S and WTE-Ts as an opportunity to produce energy from waste.

INTEGRATED SOLID WASTE MANAGEMENT SYSTEMS

In developing countries, solid waste is an issue to be managed rather than a resource to recover energy [8-10]. On the other hand, ISWM-S are being developing and spreading in the developed word [11-14]. In this research review some of the most relevant ISMW experiences are described, discussed and analysed.

For example, in the United States of America, Lee County, Florida, has one of the most advanced technologies and ISWM-S [15]. In 2011, the Resource Recovery Facility won a prestigious award, from the Solid Waste Association of America, recognising its excellent power generation system from solid waste along with its ISWM-S [16].

The Lee County WTE plant uses an incineration process generating up to 53MW of electric energy. This system meets challenging goals by not only recycling, composting but also WTE and landfilling gas utilisation for energy generation purposes [17]. In the main process of energy recovery from waste, 636 tonnes/day are incinerated at 982°C; hot gases produce steam and this powers a turbine generator. From the total energy produced, 5% is used to run plant equipment and 95% is sold to an electricity company [18]. Therefore, this ISWM-S efficiently takes advantage of all the MSW produced, generating energy from waste in addition to recycling materials and composting.

According to the Department of the Environment from the United Kingdom, in the debate about the issues regarding energy recovery from MSW, the scope is the role of anaerobic digestion and the use of landfill gas [3]. In the UK, although MSW is a current issue, landfill is still the principal management option of waste. This happens due to lower initial capital and operating costs than other options such as energy recovery, recycling and composting [19]. An Energy Recovery Facility in the UK is located in Sheffield. This plant generates electricity energy for the National Grid and heat for the District Network. This ISWM-S prevents around 21,000 tonnes of greenhouse gas emissions (GHG-E) every year [20]. This is a clear example of integrated solid waste management to produce electricity and heat.

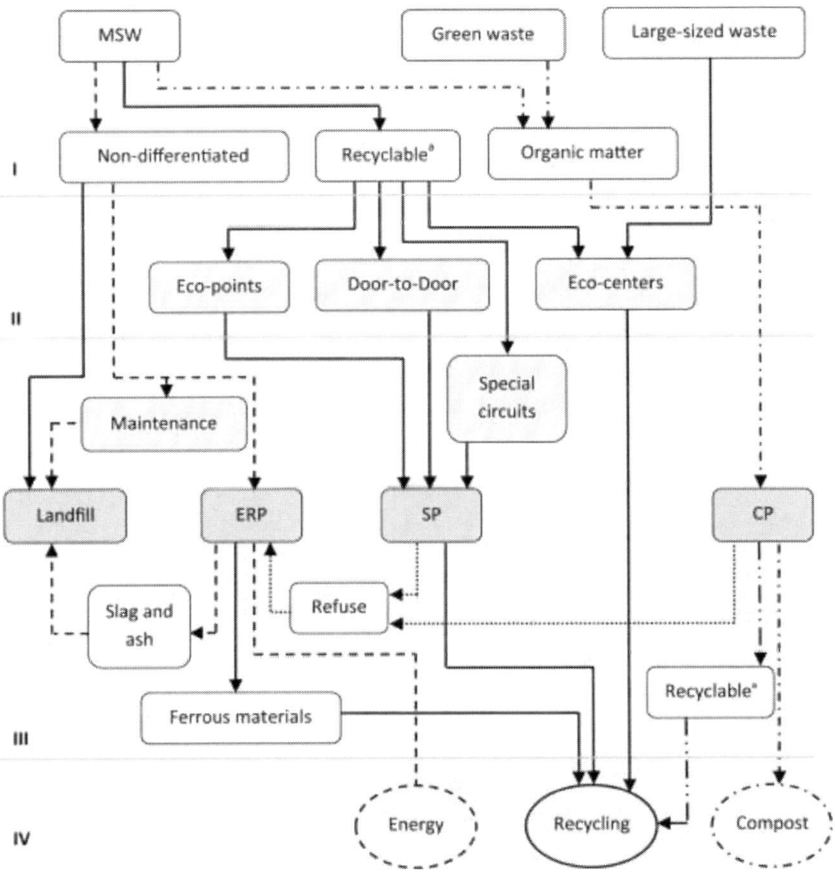

Figure 1. The ISWM-S from Porto, Portugal. ERP as the Energy Recovery Plant, SP as the Sorting Plant and CP as the Composting Plant. I – Resident: production and disposal II – Councils: production and disposal III – LIPOR: valorization and treatment IV – Clients: products and resources converted into valuable materials from municipal solid waste [21].

In Porto, Portugal, the ISWM of the Inter-municipal solid waste management system takes advangtage of the total solid waste produced in the city. This systmem has a gasification plant for organic, an incineration plant for solid waste, and a composting plant for organic fraction. The remaining pre-treated MSW is disposed into the landfill system [21]. Figure 1 shows the ISWM system from Porto. This system has four stages. In the first stage, the waste is generated by citizens. After this, the city councils collect separated waste in three different ways. In stage three, waste is treated in different facilities. Finally, energy, recycling materials and compost are the valuable products in the integrated system. In the energy recovery plant, thermal treatment produces energy. High temperatures between 1000°C and 1200°C incinerate the treated waste under excess oxygen conditions. The energetically self-sufficient plant sells 90% of the produced energy to the electricity grid network [21].

In the context of developing countries, there is a lack of the use of these techniques [4]. While engineering systems have helped industrialised countries with their SWM issues, collection and disposal are challenges in SWM in the developed word[5], involving public health problems and environmental risk [22]. While industrialised countries are implementing ISWM-S, in developing countries, these systems are limited [5]. For example, a mathematical model has been developed for a long term planning of ISWM projects including multiple variables [23], such as WTE-Ts. Conventional SWM in developing Asian countries contributes with GHG-E [24]. After considering the ISWM-S worldwide and the lessons learned from SWM systems in 222 countries, we can conclude that there is a strong relationship between waste composition and energy generation, and that the management of household solid waste contributes largely to reduce the costs from collection to energy generation [25, 26].

WASTE-TO-ENERGY TECHNOLOGIES

Around the globe, the most common techniques for managing MSW are recycling, composting, mechanical, biological treatment, and WTE-Ts [2, 3, 7, 8, 12, 25]. This chapter addresses the state-of-the-art of these technologies to recover energy from waste.

Biological Treatment of MSW to Recover Energy from Waste

The biodegradable MSW fraction has a high potential for energy production [27, 28]. Biological treatment technologies (BTT) are designed and engineered for natural biological process working with the organic rich fraction of MSW [29]. These treatments are divided into two different processes according to the conditions in which happen: the aerobic process or composting (in the presence of oxygen) and the anaerobic process (in the absence of oxygen) [30, 31]. The main product of the anaerobic process is a combustible gas which is a mixture of methane and carbon dioxide [27]. This process requires less energy than the aerobic process and creates much lower amounts of biological heat. The biodegradable fraction is converted into a fuel known as biogas. This biogas is burned to produce heat and/or electrical energy.

Anaerobic Digestion Technologies

Biogas technologies are classified as 'wet' and 'dry' techniques [32, 33]. Wet technologies process more liquid material whereas dry technologies treat drier materials. Table 1 illustrates the advanced anaerobic digestion technologies to produce biogas, their advantages and disadvantages.

In general, 'wet' anaerobic digestion technologies have been adopted in well-established systems to treat municipal wastewater. The digestion process produces biogas and decontaminated water. 'Dry' anaerobic digestion technologies operate with higher solid content and produce greater heat [29, 32]. The production of biogas reduces the amount of waste [33] and, therefore, reduces the amount of waste for disposal. This biogas is usually used in two ways: to generate electricity and to produce heat in different required processes [33, 44]. Additionally, excess heat can be used in district heating networks or in industrial processes; and future studies regarding the use of biogas as vehicle fuel are expected. In the UK, for example, biogas from anaerobic plants is injected into the national grid as a clean fuel [29].

Table 1. Anaerobic digestion technologies

Anaerobic digestion technologies	Operating Temperature	Advantages	Disadvantages	Refer-ence
Wet (low solids)	Mesophilic (35–40°C)	Used in landfills High internal rate of return Pre-treatment method to improve the efficiency of biogas process Low level of sludge generation Low operational temperatures Stable operation	Less diffusion of the technology Low investment in facilities Low government subsidies Large periods cultivations	[34-37]
	Thermophilic (55–60°C)	Production of hydrogen and methane Operation at a high organic loading rate Low operational and maintenance costs Increase gas production Resistance to foaming	Less stable - instability problems Higher residual volatile acids concentrations Limited number of digesters	[38, 39]
Dry (high solids)	Mesophilic (35°C)	Less accumulation of volatile acids Lower specific growth rate of microorganisms Highest organic matter removal rate	Lower reductions of cellulose and hemicelluloses A larger operating time to obtain methane and organic matter degradation (40 days)	[40-42]
	Thermophilic (55°C)	Greater reductions of cellulose and hemicelluloses Shorter operating time to obtain methane and organic matter degradation (20 days) Higher coefficient of methane production. Inhibited due to ammonia with organic loading rate	Accumulation of volatile fatty acids Higher specific growth rate of microorganisms	[40, 42, 43]

Thermal Treatment of MSW to Recover Energy from Waste

These technologies employ pyrolysis and gasification to recover energy from MSW [45-47]. Pyrolysis and gasification process are not new ideas [48]. This part of the study will focus on the description of pyrolysis and gasification technologies.

Pyrolysis

Pyrolysis is the thermal degradation of solid waste in the absence of oxygen [48]. External heat is required to maintain the temperature between 300 to 800°C [49], depending on the materials used in the process. In this technology, waste needs to be pre-treated. So, it requires the mechanical separation of glass, metals and inert materials [45].

This process starts with thermal decomposition of organic material, at 300°C, with reduced oxygen or oxygen-free, within heated chambers. Then, the temperature increases to 800°C in a non-reactive atmosphere; and the final bio-products of pyrolysis are gases, liquid and solid residuals [50].

Syngas is a biogas which comes from the pyrolysis process. This gas includes methane, hydrogen, carbon monoxide and carbon dioxide. The net calorific value of syngas is normally between 15 and 20 MJ/Nm^3 [51]. In addition, a recent study found that after distillation of liquid hydrocarbons (from the pyrolysis of plastic waste), the resulting synthetic product has the same properties as the petro-diesel fuel [52]. Some advantages of the pyrolysis process are: the equipment is flexible for installation; waste separation is not necessary; there are minimum environmental issues; all waste materials are used to produce different bio-products; and the produced syngas can be used in different energy applications such as engines, boilers, fuel cells, turbines and heat pumps [53-55]. In summary, the gases can be burned to produce energy, and, these gases can be condensed to produce bio-fuels.

Figure 2 illustrates an Italian case study of the pyrolysis of biomass waste to produce energy [53]. Exhaust gases are used to sustain the energy in the pyrolysis process. Thermal energy is used to produce electrical energy using a turbine and heat using a heat exchanger. It is important to clarify that pyrolysis can process biomass and plastic materials.

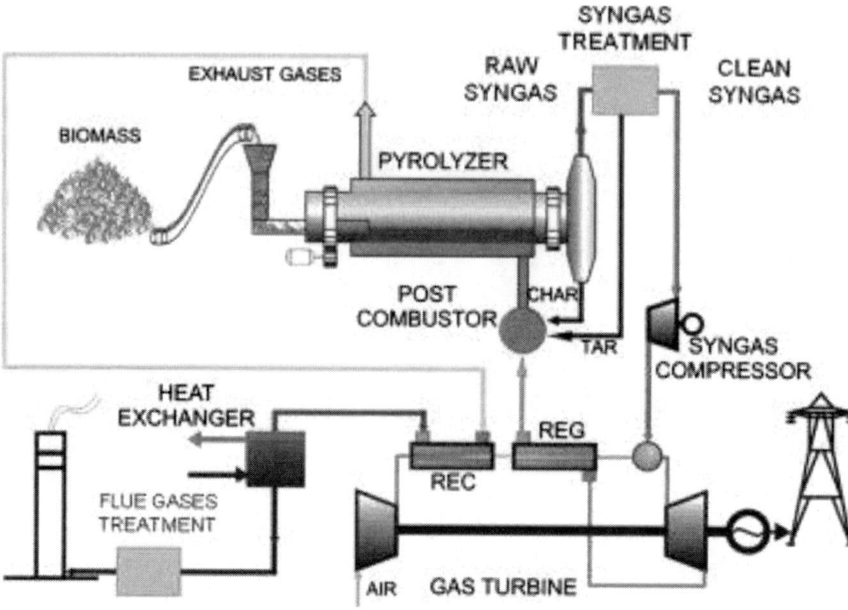

Figure 2. The Pyrolysis process [53].

Plasma pyrolysis is the highest advanced technology developed to produce syngas by transforming plastic waste with a high calorific value [56]. State-of-the-art commercial plasma technologies produce two bio-products: hydrogen-rich syngas used to generate electricity, and inert construction materials [57]. Plasma technology is a potential solution to recover energy from solid waste.

Gasification

The Gasification process to produce energy from waste has been developed over the last three decades [58]. This bioprocess involves partial oxidation and its main product is fuel gas [59]. Gasification can reduce the waste mass by 70% and over 90% of the volume of waste [60]. A recent study indicated that this thermal treatment of waste is a practical viable option for WTE conversion, limiting greenhouse emissions and reducing landfill disposal options [61]. In this process, part of the energy is used to produce heat to gasify the remaining fuel [60].

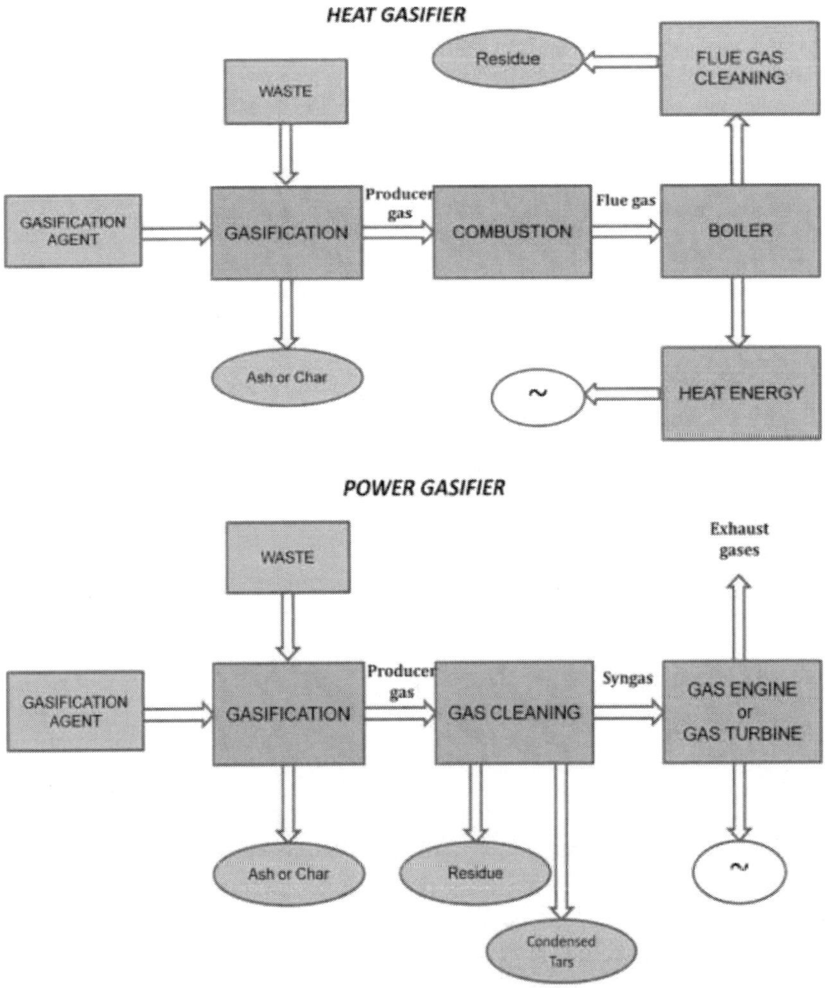

Figure 3. Representation of the gasification technology based on configurations for heat and power [61].

The main advantages of WTE-T are: reduction of organic contaminants; reduction of mass and volume of waste (80% and 90% respectively); high potential for the saving of land; use of recyclables; reduction of emissions and environmental burdens; environmental compatible for co-generation (heat and electricity production) as a renewable resource of waste; technical and economic feasibility; and it has a high range of working temperatures (700-900°C) [58, 60-62].

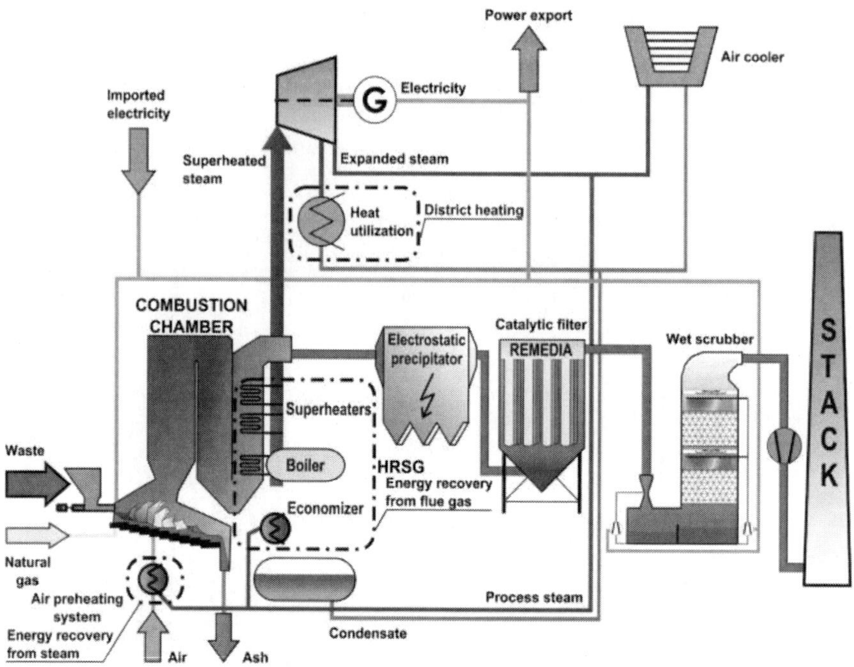

Figure 4. Incineration plant to produce electricity and heat from MSW [64].

In general terms, the biogas (syngas) produced from gasification process is energy-rich and clean enough to produce energy in a gas turbine or engine [59]. However, nowadays, there are eight different reactor technologiesfor the waste gasification process (rotary kiln; fixed, entrained and fluidized bed; plasma reactor; vertical shaft and moving grate furnace) and different plan configurations [61]. The syngas fuel is a mixture of hydrogen, carbon dioxide, methane and hydrogen which is a potential source to produce electricity in a turbine-based power plant [63]. To summarise, Figure 3 shows the gasification WTE-T power plant to produce energy and heat.

Incineration Treatment of MSW to Recovery Energy from Waste

Figure 4 illustrates a classic incineration WTE-T. First, waste is directly burned in the combustion chamber at an adequate temperature (900-950°C) using flue gas and pre-heated air. After the waste incineration process, superheated steam is produced and then it is used within a cogeneration system to produce energy and heat. The electric energy is produced by a turbine connected to a generator and the heat by a district heating system [64].

The highest environmental impact of MSW incineration is the production of GHG-E causing public health concerns [65, 66]. However, some efforts to control these concerns have resulting in regulations to reduce the amount of specific emissions such as: sulphur dioxide, hydrogen fluoride, nitrogen oxide and nitrogen dioxide, total organic carbon, hydrogen chloride, carbon monoxide, dust and volatile heavy metal [67]. On the other hand, some advanced technologies have been developed to reduce these effects. For example, a hybrid plant using incineration and gasification has been developed to reduce volatile heavy metals [66].

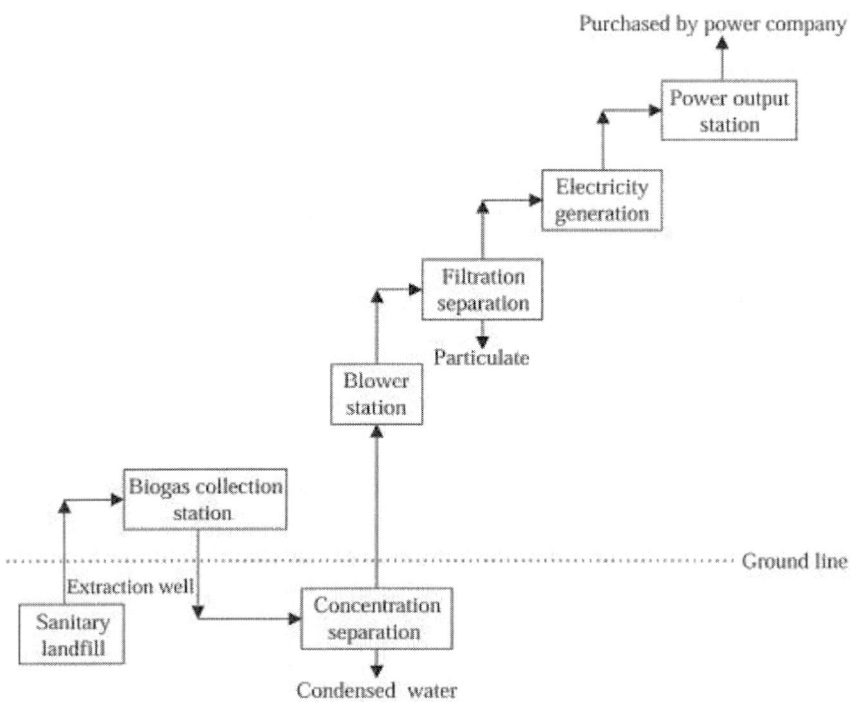

Figure 5. Flow diagram of electricity generation from landfill biogas [72].

Landfill Gas Utilization

Gas emissions from municipal landfills around the world are causing global environmental impacts [68]. This biogas is rich in methane which must be used to produce energy and heat in cogeneration plants [69]. A recent study showed that worldwide landfills produce about 75 billion Nm3 and less than 3% of this potential is used to produce energy or heat. In the USA, for example, modern regulated landfills collect approximately 2.6 million tons of methane producing heat and electricity with a capacity up to 50 MW turbine generator [70].

Landfill gas to produce electricity can use different innovative energy system technologies such as gas turbine in Brayton cycle, Organic Rankine Cycle, Stirling cycle engine or Solid Oxide fuel cells [71]. Figure 5 shows the stages to produce electricity from landfill gas [72]. The appropriate treatment to landfills will result in land space remediation and reduction of GHG-E reducing public health issues [68, 73].

Analysis of the Use WTE-T as an ISWM

From the reviewed literature, it is clear that WTE-Ts in developed countries have been developed to a large scale. However, not all environmental issues have been dealt with. For example, Germany and the USA have ISWM-S (from collection to energy generation and disposal) whereas in the UK, these technologies recently are being discussed and implemented [3, 7, 14]. The most important issue in the global knowledge system is that these technologies are privatised and, even between developed countries, information technologies is not shared in the interest of environmental problems, energy security and public health concerns generated by inefficient solid waste management systems.

In developing countries, the scenario is even worse. Collection, transportation and disposal of solid waste are the current challenges [4, 8]. However, developed countries are the highest contributors with GHG-E to the environment [2]. Whereas developing countries produce environmental goods for free, developed countries produce knowledge which is sold to both developed and developing countries. WTW-Ts in developing countries need to be implemented in order to maximise the environmental goods in the world, reduce poverty and increase energy production from waste.

CONCLUSION

The municipal solid waste is a potential energy source in both developed and developing countries. However, clearly, there is a lack of technology transfer between these countries. For example, whereas the USA has one the highest developed ISWM-S in the world, countries such as India and Nepal has relevant issues in the collection, transport and disposal of MSW.

Currently, there are a large number of technologies to produce energy from MSW. Biogases from anaerobic bio-digestion process can produce electricity whereas direct incineration of waste might produce steam and then produce energy using turbines and generators. However, gas utilisation from landfills is a potential source to produce renewable energy which has not yet been exploited. Finally, ISWM-S is a solution, along with energy recovery from waste, to manage MSW in an effective and efficient way.

REFERENCES

[1] D. Hoornweg and P. Bhada-Tata, *"What a waste: a global review of solid waste management,"* 2012.

[2] H. Cheng and Y. Hu, "Municipal solid waste (MSW) as a renewable source of energy: Current and future practices in China," *Bioresource Technology,* vol. 101, pp. 3816-3824, June 2010.

[3] Department for Environment Food and Rural Affairs from UK. (2014, May 27). *Energy from waste – a guide to the debate*. Available: https://www.gov.uk/government/uploads/system/uploads/attachment_data/file/284612/pb14130-energy-waste-201402.pdf.

[4] L. A. Guerrero, G. Maas, and W. Hogland, "Solid waste management challenges for cities in developing countries," *Waste Management,* vol. 33, pp. 220-232, January 2013.

[5] R. E. Marshall and K. Farahbakhsh, "Systems approaches to integrated solid waste management in developing countries," *Waste Management,* vol. 33, pp. 988-1003, April 2013.

[6] J. D. Nixon, P. K. Dey, S. K. Ghosh, and P. A. Davies, "Evaluation of options for energy recovery from municipal solid waste in India using the hierarchical analytical network process," *Energy,* vol. 59, pp. 215-223, September 15 2013.

[7] C. Psomopoulos, A. Bourka, and N. Themelis, "Waste-to-energy: a review of the status and benefits in USA," *Waste Management,* vol. 29, pp. 1718-1724, May 2009.

[8] K. Kalyani and K. Pandey, "Waste to energy status in India: a short review," *Renewable and Sustainable Energy Reviews,* vol. 31, pp. 113-120, March 2014.

[9] D. Pokhrel and T. Viraraghavan, "Municipal solid waste management in Nepal: practices and challenges," *Waste Management,* vol. 25, pp. 555-562, March 2005.

[10] R. Singh, V. Tyagi, T. Allen, M. Ibrahim, and R. Kothari, "An overview for exploring the possibilities of energy generation from municipal solid waste (MSW) in Indian scenario," *Renewable and Sustainable Energy Reviews,* vol. 15, pp. 4797-4808, December 2011.

[11] S. Kathirvale, M. Muhd Yunus, K. Sopian, and A. Samsuddin, "Energy potential from municipal solid waste in Malaysia," *Renewable Energy,* vol. 29, pp. 559-567, April 2004.

[12] M. Münster, "Energy System Analysis of Waste-to-Energy technologies," PhD, Department of Development and Planning Aalborg University, Aalborg University, Aalborg, Denmark, 2009.

[13] M. Münster and H. Lund, "Comparing Waste-to-Energy technologies by applying energy system analysis," *Waste Management,* vol. 30, pp. 1251-1263, July 2010.

[14] M. Münster and P. Meibom, "Optimization of use of waste in the future energy system," *Energy,* vol. 36, pp. 1612-1622, March 2011.

[15] H. Drinc. (2011, 19th May). *Lee County Waste-to-Energy and Recovered Material Processing Facility Expansions.* Available: http://www.hdrinc.com/portfolio/lee-county-waste-to-energy-and-recovered-material-processing-facility-expansions.

[16] Lee County Government Southwest Florida. (2014, May 27). *Solid Waste Division Program recognitions & awards.* Available: http://www.leegov.com/gov/dept/SolidWaste/Facilities/Pages/WasteToEnergyAwardsreceived.aspx.

[17] Lee County Government Southwest Florida. (2014, May 27). *Waste-To-Energy: How It Works.* Available: http://www.leegov.com/gov/dept/SolidWaste/Facilities/Pages/WssteToEnergyHowitWorks.aspx.

[18] K. Howard. (2009, May 27). *Integrated Solid Waste Systems provided by the Lee County Solid Waste Division.* Available: http://lee.ifas.ufl.edu/AgNatRes/Sustainability/SustainableInfo/IntegratedSolidWaste-Sustainability_Conference%20FM_Oct2010.pdf.

[19] P. Pukrittayakamee, *"Community-scale Integrated Solid Waste Management System."* Master of Science in Renewable Energy Systems and the Environment Research, Department of Mechanical Engineering University of Strathclyde, UK, 2010.

[20] Veolia. (2014, May 25). *Energy Recovery Facility*. Available: http://www.veoliaenvironmentalservices.co.uk/Sheffield/What-happens-to-your-waste/Energy-Recovery-Facility/Energy-Recovery--The-Process/

[21] M. Herva, B. Neto, and E. Roca, "Environmental assessment of the integrated municipal solid waste management system in Porto (Portugal)," *Journal of Cleaner Production,* vol. 70, pp. 183-193, May 1 2014.

[22] L. Rodic. (2011, May 27). *Comparing Solid Waste Management in the World's Cities*. Available: http://unpan1.un.org/intradoc/groups/public/documents/apcity/unpan050309.pdf.

[23] A. K. Srivastava and A. K. Nema, "Fuzzy parametric programming model for multi-objective integrated solid waste management under uncertainty," *Expert Systems with Applications,* vol. 39, pp. 4657-4678, April 2012.

[24] S. N. Othman, Z. Zainon Noor, A. H. Abba, R. O. Yusuf, and M. A. Abu Hassan, "Review on life cycle assessment of integrated solid waste management in some Asian countries," *Journal of Cleaner Production,* vol. 41, pp. 251-262, February 2013.

[25] A. Laurent, I. Bakas, J. Clavreul, A. Bernstad, M. Niero, E. Gentil, et al., "Review of LCA studies of solid waste management systems – Part I: Lessons learned and perspectives," *Waste Management,* vol. 34, pp. 573-588, March 2014.

[26] A. Laurent, J. Clavreul, A. Bernstad, I. Bakas, M. Niero, E. Gentil, et al., "Review of LCA studies of solid waste management systems – Part II: Methodological guidance for a better practice," *Waste Management,* vol. 34, pp. 589-606, March 2014.

[27] V. Nallathambi Gunaseelan, "Anaerobic digestion of biomass for methane production: A review," *Biomass and Bioenergy,* vol. 13, pp. 83-114, March 1997.

[28] D. L. Wise, R. G. Kispert, and E. W. Langton, "A review of bioconversion systems for energy recovery from municipal solid waste Part II: Fuel gas production," *Resources and Conservation,* vol. 6, pp. 117-136, August 1981.

[29] Department for Environment Food and Rural Affairs from UK. (2014, May 27). *Advanced Biological Treatment of Municipal Solid Waste*. Available: https://www.gov.uk/government/uploads/system/uploads/attachment_data/file/284612/pb14130-energy-waste-201402.pdf.

[30] K. Fricke, H. Santen, and R. Wallmann, "Comparison of selected aerobic and anaerobic procedures for MSW treatment," *Waste Management,* vol. 25, pp. 799-810, October 2005.

[31] A. J. Ward, P. J. Hobbs, P. J. Holliman, and D. L. Jones, "Optimisation of the anaerobic digestion of agricultural resources," *Bioresource Technology,* vol. 99, pp. 7928-7940, November 2008.

[32] B. Scaglia, S. Salati, A. Di Gregorio, A. Carrera, F. Tambone, and F. Adani, "Short mechanical biological treatment of municipal solid waste allows landfill impact reduction saving waste energy content," *Bioresource Technology,* vol. 143, pp. 131-138, September 2013.

[33] B. Zhu, P. Gikas, R. Zhang, J. Lord, B. Jenkins, and X. Li, "Characteristics and biogas production potential of municipal solid wastes pretreated with a rotary drum reactor," *Bioresource Technology,* vol. 100, pp. 1122-1129, February 2009.

[34] K. Hii, S. Baroutian, R. Parthasarathy, D. J. Gapes, and N. Eshtiaghi, "A review of wet air oxidation and Thermal Hydrolysis technologies in sludge treatment," *Bioresource Technology,* vol. 155, pp. 289-299, March 2014.

[35] S. N. Malik, T. Saratchandra, P. D. Tembhekar, K. V. Padoley, S. L. Mudliar, and S. N. Mudliar, "Wet air oxidation induced enhanced biodegradability of distillery effluent," *Journal of Environmental Management,* vol. 136, pp. 132-138, April 1 2014.

[36] S. Supaphol, S. N. Jenkins, P. Intomo, I. S. Waite, and A. G. O'Donnell, "Microbial community dynamics in mesophilic anaerobic co-digestion of mixed waste," *Bioresource Technology,* vol. 102, pp. 4021-4027, March 2011.

[37] L. Yan, Y. Gao, Y. Wang, Q. Liu, Z. Sun, B. Fu, et al., "Diversity of a mesophilic lignocellulolytic microbial consortium which is useful for enhancement of biogas production," *Bioresource Technology,* vol. 111, pp. 49-54, May 2012.

[38] S. Ghanimeh, M. El Fadel, and P. Saikaly, "Mixing effect on thermophilic anaerobic digestion of source-sorted organic fraction of municipal solid waste," *Bioresource Technology,* vol. 117, pp. 63-71, August 2012.

[39] K. Stamatelatou, G. Antonopoulou, and P. Michailides, "15 - Biomethane and biohydrogen production via anaerobic digestion/fermentation," in *Advances in Biorefineries*, K. Waldron, Ed., ed: Woodhead Publishing, 2014, pp. 476-524.

[40] J. Fernández-Rodríguez, M. Pérez, and L. I. Romero, "Comparison of mesophilic and thermophilic dry anaerobic digestion of OFMSW: Kinetic analysis," *Chemical Engineering Journal,* vol. 232, pp. 59-64, October 2013.

[41] A. Giuliano, D. Bolzonella, P. Pavan, C. Cavinato, and F. Cecchi, "Co-digestion of livestock effluents, energy crops and agro-waste: Feeding and process optimization in mesophilic and thermophilic conditions," *Bioresource Technology,* vol. 128, pp. 612-618, January 2013.

[42] J. Shi, Z. Wang, J. A. Stiverson, Z. Yu, and Y. Li, "Reactor performance and microbial community dynamics during solid-state anaerobic digestion of corn stover at mesophilic and thermophilic conditions," *Bioresource Technology,* vol. 136, pp. 574-581, May 2013.

[43] H. V. Kinnunen, P. E. P. Koskinen, and J. Rintala, "Mesophilic and thermophilic anaerobic laboratory-scale digestion of Nannochloropsis microalga residues," *Bioresource Technology,* vol. 155, pp. 314-322, March 2014.

[44] R. Bayard, J. de Araújo Morais, G. Ducom, F. Achour, M. Rouez, and R. Gourdon, "Assessment of the effectiveness of an industrial unit of mechanical–biological treatment of municipal solid waste," *Journal of Hazardous Materials,* vol. 175, pp. 23-32, March 15 2010.

[45] Department for Environment Food and Rural Affairs from UK. (2014, May 28). *Advanced Thermal Treatment of Municipal Solid Waste.* Available: https://www.gov.uk/government/uploads/system/uploads/attachment_data/file/221035/pb13888-thermal-treatment-waste.pdf.

[46] P. Lettieri, L. Yassin, and S. J. R. Simons, "7 - Advanced thermal treatment of composite wastes for energy recovery," in *Management, Recycling and Reuse of Waste Composites*, V. Goodship, Ed., ed: Woodhead Publishing, 2010, pp. 152-191.

[47] A. Tabasová, J. Kropáč, V. Kermes, A. Nemet, and P. Stehlík, "Waste-to-energy technologies: Impact on environment," *Energy,* vol. 44, pp. 146-155, August 2012.

[48] U. Arena and F. Di Gregorio, "Element partitioning in combustion- and gasification-based waste-to-energy units," *Waste Management,* vol. 33, pp. 1142-1150, May 2013.

[49] M. Agarwal, J. Tardio, and S. Venkata Mohan, "Critical analysis of pyrolysis process with cellulosic based municipal waste as renewable source in energy and technical perspective," *Bioresource Technology,* vol. 147, pp. 361-368, November 2013.
[50] R. C. Newbold. (2014, May 28). *Pyrolysis (Thermolysis) as a Preferred BioSynENERGY Technology for Generating Value from Waste.* Available: http://www.biosynenergy.com/Pyrolysis-Technology.php.
[51] S. Zafar. (2014, May 28). *Overview of Biomass Pyrolysis*. Available: http://www.bioenergyconsult.com/tag/biofuels/
[52] B. K. Sharma, B. R. Moser, K. E. Vermillion, K. M. Doll, and N. Rajagopalan, "Production, characterization and fuel properties of alternative diesel fuel from pyrolysis of waste plastic grocery bags," *Fuel Processing Technology,* vol. 122, pp. 79-90, June 2014.
[53] B. D'Alessandro, M. D'Amico, U. Desideri, and F. Fantozzi, "The IPRP (Integrated Pyrolysis Regenerated Plant) technology: From concept to demonstration," *Applied Energy,* vol. 101, pp. 423-431, January 2013.
[54] T. Malkow, "Novel and innovative pyrolysis and gasification technologies for energy efficient and environmentally sound MSW disposal," *Waste Management,* vol. 24, pp. 53-79, February 2004.
[55] P. T. Williams, "Pyrolysis of waste tyres: A review," *Waste Management,* vol. 33, pp. 1714-1728, August 2013.
[56] M. Punčochář, B. Ruj, and P. K. Chatterj, "Development of Process for Disposal of Plastic Waste Using Plasma Pyrolysis Technology and Option for Energy Recovery," *Procedia Engineering,* vol. 42, pp. 420-430, 2012.
[57] P. Moayedi. (2014, May 28). *Gasplasma* Available: http://www.advancedplasmapower.com/about-us/our-company/
[58] M. Morris and L. Waldheim, "Energy recovery from solid waste fuels using advanced gasification technology," *Waste Management,* vol. 18, pp. 557-564, October 1998.
[59] T. Li, R. Grace, J. Lim, P. Watkinson, P. Chen, and R. Kim, "Biomass gasification in a circulating fluidized bed," *Biomass and Bioenergy,* vol. 26, pp. 171-193, February 2004.
[60] U. Arena, "Gasification: An alternative solution for waste treatment with energy recovery," *Waste Management,* vol. 31, pp. 405-406, March 2011.
[61] U. Arena, "Process and technological aspects of municipal solid waste gasification. A review," *Waste Management,* vol. 32, pp. 625-639, April 2012.

[62] U. Arena, F. Di Gregorio, C. Amorese, and M. L. Mastellone, "A techno-economic comparison of fluidized bed gasification of two mixed plastic wastes," *Waste Management,* vol. 31, pp. 1494-1504, July 2011.

[63] H. A. Arafat and K. Jijakli, "Modeling and comparative assessment of municipal solid waste gasification for energy production," *Waste Management,* vol. 33, pp. 1704-1713, August 2013.

[64] B. Kilkovsky, P. Stehlik, Z. Jegla, L. Tovazhnyansky, O. Arsenyeva, and P. Kapustenko, "Heat exchangers for energy recovery in waste and biomass to energy technologies – I. Energy recovery from flue gas," *Applied Thermal Engineering,* vol. 64, pp. 213-223, March 2014.

[65] A. Beylot and J. Villeneuve, "Environmental impacts of residual Municipal Solid Waste incineration: A comparison of 110 French incinerators using a life cycle approach," *Waste Management,* vol. 33, pp. 2781-2788, 12//2013.

[66] M. Wu, C. Lin, and W. Zeng, "Effect of waste incineration and gasification processes on heavy metal distribution," *Fuel Processing Technology,* vol. 125, pp. 67-72, 9//2014.

[67] Department for Environment Food and Rural Affairs from UK. (2014, May 27). *Incineration of Municipal Solid Waste.* Available: https://www.gov.uk/government/uploads/system/uploads/attachment_data/file/221036/pb13889-incineration-municipal-waste.pdf.

[68] A. Niskanen, H. Värri, J. Havukainen, V. Uusitalo, and M. Horttanainen, "Enhancing landfill gas recovery," *Journal of Cleaner Production,* vol. 55, pp. 67-71, 9/15/2013.

[69] P. Mostbauer, L. Lombardi, T. Olivieri, and S. Lenz, "Pilot scale evaluation of the BABIU process – Upgrading of landfill gas or biogas with the use of MSWI bottom ash," *Waste Management,* vol. 34, pp. 125-133, 1//2014.

[70] N. J. Themelis and P. A. Ulloa, "Methane generation in landfills," *Renewable Energy,* vol. 32, pp. 1243-1257, 6//2007.

[71] R. Bove and P. Lunghi, "Electric power generation from landfill gas using traditional and innovative technologies," *Energy Conversion and Management,* vol. 47, pp. 1391-1401, 7//2006.

[72] W. T. Tsai, "Bioenergy from landfill gas (LFG) in Taiwan," *Renewable and Sustainable Energy Reviews,* vol. 11, pp. 331-344, 2//2007.

[73] P. Frändegård, J. Krook, N. Svensson, and M. Eklund, "A novel approach for environmental evaluation of landfill mining," *Journal of Cleaner Production,* vol. 55, pp. 24-34, September 15 2013.

In: Energy Efficiency
Editor: Julian E. Perez

ISBN: 978-1-53611-040-1

Chapter 3

ENERGY-EFFICIENT POWER ALLOCATION WITH HARD PROPORTIONAL RATE GUARANTEE IN OFDMA SYSTEMS

Mau-Luen Tham[1], PhD, and Su Fong Chien[2], PhD

[1]Department of Electrical and Electronic Engineering,
Lee Kong Chian Faculty of Engineering and Science,
Universiti Tunku Abdul Rahman (UTAR), Kajang, Malaysia
[2]Department of Strategic Advanced Research (StAR),
MIMOS Berhad, Kuala Lumpur, Malaysia

ABSTRACT

Orthogonal frequency division multiple access (OFDMA), the air interface for major and future wireless communication systems, can achieve significant energy saving while satisfying various quality-of-service (QoS) requirements with intelligent management of radio resources such as subcarriers and transmission power. One of the common QoS metrics belongs to proportional fairness, which enforces the achieved data rate proportions among users. Together with this set of nonlinear equality constraints, the problem of energy-efficient resource allocation can be modeled as a mixed-integer fractional program, which is computationally difficult to solve. Separating the subcarrier and power allocation is a prevalently adopted alleviation to reduce complexity while approaching the optimal system performance. Under a pre-determined set of subcarrier allocation, most existing solutions aim to maximize the energy efficiency of OFDMA systems, which belongs to nonlinear fractional program. However, their derived solutions are based on some ideal assumptions such as high channel conditions. When channel conditions degrade, the adherence to the proportionality constraints will no longer be strictly obeyed. In this chapter, our goal is to maximize the energy efficiency of OFDMA systems subject to the constraints on base station transmission power and proportional data rates. We first convert the nonlinear fractional problem into a simpler single-variable nonlinear problem via Charnes-Cooper Transformation. By exploring the structure of the new problem, we then develop an iterative algorithm to find the optimal solution of the considered problem while delivering hard guarantee of the proportional data rates. Simulation results reveal that our proposed scheme satisfies the rate proportionality constraint in the strictest sense, as compared to previous algorithm.

Keywords: energy efficiency, OFDMA, power allocation, proportional rate constraints

1. INTRODUCTION

The proliferation of mobile terminals has accelerated the demand for network capacity. To support manifold capacity growth, advanced wireless physical-layer techniques such as multiple-input-multiple-output (MIMO) and orthogonal frequency division multiplexing (OFDM) have been developed. An advanced version of OFDM, referred to as orthogonal frequency division multiple access (OFDMA), enables multiple users to share the radio resources simultaneously. Due to its high speed capacity, OFDMA has been widely adopted in all major wireless communication systems such as Long Term Evolution (LTE). Since wireless resources (e.g., transmission power and subcarriers) are inherently limited, how to efficiently allocate them constitutes a key challenge. The common objective in resource allocation is to enhance the spectral efficiency (SE) [17]. This, however, has raised high energy costs and serious environmental issues [18]. For example, [19] reports that the telecommunication industry accounts for two percent of the total CO_2 emissions worldwide. Accordingly, there has been a paradigm shift from increasing SE to improving energy efficiency (EE).

Energy efficient designs of OFDMA-based base stations (BSs) have become a central topic in the field of green communications. This is contributed by the fact that BS represents the main source of energy consumption in cellular networks. Another fueling factor is that the growing interest in new mobile services has resulted in an increased number of installed BSs worldwide [1]. Because EE is often measured in "bits per Joule per Hz" which is linked to the ratio of Shannon capacity to power consumption at the electrical circuits, energy-efficient optimization belongs to a class of fractional programming. Various numerical techniques such as parameterized convex programming [2], Charnes-Cooper transformation (CCT) [3], and bi-level optimization [4] have been proposed to solve the EE problem. However, the focus is restricted to a single-user framework. Works that do consider a multiuser scenario can be found in [5-7].

Since the requirements for the future services continue to evolve, it is crucial that the developed energy-efficient techniques are flexible enough to handle as high data rates as possible with various quality-of-service (QoS) requirements. Generally speaking, the QoS constraints of these schemes fall into one of two categories: minimum user data rates [2, 4, 6, 7, 8] and proportional data rates [5, 9]. In the first category, mobile operators have to entertain each individual user data rate requirements, while maximizing the EE. However, forcing the BS to satisfy the minimum rate requirements will degrade the EE since a power allocation that maximizes the EE will yield relatively low transmission rates. The EE degradation becomes much more severe when most of mobile subscribers have poor communication link quality. Another concern arising from this strategy is the infeasibility of the solution due to limited BS transmission power [8]. In the second category, the schemes address the preceding issues by letting users to accept proportional rates, in which high-priority services can have high data rates. In this way, network operators have the flexibility to control the capacity ratios among users of different service priorities while ensuring that the solution is always reliable. However, these schemes typically adopt the Lagrangian Dual method to solve their energy-efficient problems under the assumption that the duality gap between the original problem and dual problem is negligible. In fact, there exists potential inaccuracy caused by the duality gap found in the Lagrangian Dual method [8]. This, in turn, makes the solution found in the dual domain unable to meet all the QoS constraints accurately [7]. In [9], numerical methods are employed to solve the EE problem, however, the performance analysis for adherence to rate proportionality is not presented. Therefore, it is not clear, how strict the achieved user rates gets to that pre-determined rate proportionality constant. Since the legitimate judges of service satisfaction level are end users, the accuracy in provisioning QoS requirements is important to operators, which deserves further study.

In this chapter, we extend the work in [10] and aim at maximizing the EE of OFDMA systems under the constraints on transmission power and proportional data rates. Different from [10], the proposed scheme provides hard guarantee instead of soft guarantee of rate proportionality. This is accomplished by solving a derived single-variable nonlinear equation until all the remaining subcarriers are assigned with positive power levels. In this way, the proposed scheme satisfies the QoS constraints in the strictest sense regardless of channel conditions. Furthermore, the proposed scheme can operate in one of two EE modes namely optimal-power mode and full-power mode. In the first mode, the BS consumes the least amount of transmission power and thus produces the highest EE. This can be useful when the user density is low. However, when the network is congested, the BS can enter the second mode to increase transmission capacity. The rest of the chapter is organized as follows. Section 2 describes the system model. Section 3 presents the proposed resource allocation algorithm. Section 4 gives simulation results. Conclusions are drawn in Section 5.

2. System Model

Considering a downlink OFDMA system serving a set of mobile users K = $\{1,2,\ldots,K\}$, each with different QoS requirements. All users share a set of subcarriers N = $\{1,2,\ldots,N\}$ with each subcarrier occupying a bandwidth of $B_n = B/N$, where B_n is assumed to be smaller than coherence bandwidth and B denotes the entire system bandwidth. At the BS, the scheduler allocates transmission power $p_{k,n}$ to subcarrier n, which is assigned exclusively to one of K users. User data is modulated onto its own subcarrier(s) in units of bits and subsequently formed into an OFDMA symbol by using the inverse fast Fourier transform (IFFT). The signal is then transmitted through a slowly time-varying, frequency-selective Rayleigh channel, which can be divided into N flat fading channels. At user k, the received signal $\Upsilon_{k,n}$ for subcarrier n's data symbol $x_{k,n}$ is written as

$$\Upsilon_{k,n} = x_{k,n}\sqrt{p_{k,n}}\,g_{k,n} + \eta_n \tag{1}$$

where η_n denotes as the additive white Gaussian noise (AWGN) of power spectral density N_0 with mean zero and variance $\sigma^2 = N_0 B_n$. $g_{k,n}$ represents the magnitude of channel fading for subcarrier n between the base station and user k. It captures the effect of path loss, shadowing, frequency response and Doppler shift. Note that the fading coefficients of different users are mutually independent, which is called multiuser diversity. The slowly time-varying characteristic enables the scheduler to reconfigure resource allocation settings in a symbol-by-symbol manner, assuming that the channel gain $G_{k,n} = |g_{k,n}|^2$ remains invariant within each symbol, but can vary from symbol to another. Armed with the perfect channel information, the achievable data rate of subcarrier n of user k can be expressed as

$$r_{k,n} = \log_2(1 + p_{k,n} H_{k,n}) \text{ bits/s/Hz} \tag{2}$$

where $H_{k,n} - |g_{k,n}|^2/\sigma^2$ is known as channel-to-noise ratio (CNR).

The objective is to maximize the EE of the OFDMA system while satisfying the total transmission power P_T and QoS constraints. As in [11], the QoS constraints are defined as proportional rate constraints, i.e., the data rate of each user should adhere to a set of pre-determined proportionality constants. The benefits of introducing proportional fairness into the system are threefold. First, from the perspective of network operators, the ability to control the capacity ratios among users with a wide variety of subscription levels is an integral part of mobile services. For example, in a real cellular network such as LTE, each mobile application (or service) is assigned one QoS Class Identifier (QCI) with either guaranteed bit rate (GBR) or non-GBR [12]. Given a sufficient total available transmission power, each service can be provisioned with its target data rate. Second, in the context of green communication systems, offering proportional rates instead of minimum rates to mobile subscribers will yield higher energy efficiency. Third, in the scenario of network congestion, the proportional rate constraints can be mapped to proportional delay constraints [5].

The optimization problem can be mathematically formulated as

$$\max_{p_{k,n}, s_{k,n}} \; EE = \frac{\sum_{k=1}^{K}\sum_{n=1}^{N} s_{k,n}\log_2(1+p_{k,n}H_{k,n})}{\tau s_{k,n}\sum_{n=1}^{N} p_{k,n} + P_c} \tag{3a}$$

$$\text{s. t. } p_{k,n} \geq 0 \; \forall k \in \mathrm{K}, n \in \mathrm{N} \tag{3b}$$

$$\sum_{k=1}^{K} p_{k,n} \leq P_T \tag{3c}$$

$$p_{k,n} = \{0,1\} \; \forall k \in \mathrm{K}, n \in \mathrm{N} \tag{3d}$$

$$\sum_{k=1}^{K} p_{k,n} = 1 \ \forall n \in \mathrm{N} \tag{3e}$$

$$\alpha_m R_1 - \alpha_1 R_m = 0 \ \forall m = 2,3, \dots, K \tag{3f}$$

The performance metric of EE is reflected by the objective function in (3a), which is a ratio of transmission rate to total power consumption. τ is the inverse of power amplifier efficiency and P_c represents the circuit power consumed by signal processing and battery backup. Constraint (3b) imposes the positivity on power levels. Constraint (3c) implies that the total transmission power must be equal or less than the total available power. In constraint (3d), $s_{k,n}$ can be only either 0 or 1, indicating whether subcarrier n is used by user k or not. Constraint (3e) implies that each subcarrier can only be used by one user. Finally, $\{\alpha_1, \alpha_2, \dots, \alpha_K\}$ is a set of pre-determined parameters that are adopted to enforce proportional fairness among users, where the data rate for user k, denoted as R_k, is defined as

$$R_k = \sum_{k=1}^{K} \sum_{n=1}^{N} s_{k,n} \log_2\left(1 + p_{k,n} H_{k,n}\right) \tag{4}$$

A larger value generally means higher priority in terms of service level. When all α_k have the same value, the system offers same transmission capacity to all users. Here, the spectral efficiency (SE) is equal to $\sum_{k=1}^{K} R_k$.

3. The Proposed Energy-Efficient Resource Allocation

The optimization problem in (3) belongs to nonconvex optimization problem due to the discrete set of subcarrier assignment. Hence, it is prohibitively complex to obtain the optimal solution $\{p_{k,n}^{*}, s_{k,n}^{*}\}$ since there are $(K)^N$ possible subcarrier assignment, especially for a large-scale OFDMA system. The impracticality of optimal resource allocation makes suboptimal algorithms with lower computational complexity the preferred solution. It has been shown in [11] that separating the subchannel and power allocation is an effective way to lessen the complexity while approaching the optimal system performance. Therefore, in order to simplify the solution development and analysis of the resource allocation scheme, the energy-efficient resource allocation is segmented into two steps: subcarrier assignment and power allocation.

3.1. Subcarrier Assignment

In this section, we discuss a suboptimal subcarrier algorithm based on [10]. In the suboptimal subcarrier algorithm, it is assumed that the total available power is equally distributed to all subcarriers. The set of subcarriers assigned to user k is denoted as Ω_k. The algorithm can then be described as

1. Initialization: Set $R_k = 0, \Omega_k = \emptyset \; \forall k \in \mathrm{K}$ and $p = P_T/N$
2. First round subcarrier assignment: Greedy assignment
   - Assign a subcarrier associated with the best CNR to each of the corresponding user, $\forall k \in \mathrm{K}$
   - Record the highest gain subchannel $h_{k,1}$ for each $\Omega_k \; \forall k \in \mathrm{K}$
   - Compute each data rate R_k and let $R = \sum_{k=1}^{K} R_k$
   - Set the denominator in (3) as $\tau N p + P_c$ and define $P_w = \tau N p$
   - Compute EE
3. Assign the rest of the unassigned subchannels to each user with the conditions below:
   - Determine a user m with the least R_k/α_k
   - For user m, compute $\Delta H_{m,n} = 1/H_{m,1} - 1/H_{m,n}$
   - Minus P_w with τp and replace by τp_n, where $p_n = p + 1/H_{m,1} - 1/H_{m,n}$
   - Set $EE_{temp} = R + \log_2\left(1 + p_{m,n} H_{m,n}\right)$
   - If $H_{m,n}$ is free and $EE_{temp} - EE > 0$, then subcarrier H_n is assigned, if not go to next step
   - Find a subcarrier with $\min\left|EE_{temp} - EE\right|$ and assign this subcarrier to user m
   - Update R, P_w and EE. Go back to 3 until no subcarriers left.

Motivated by the fact that the EE is directly proportional to channel quality, the principle of the suboptimal subcarrier method is to advocate each user to use the subcarriers with high CNR as much as possible. The computational complexity is $O(NK)$.

3.2. Power Allocation

Given a predetermined subcarrier allocation, the optimization problem in (3) can be reformulated as

$$\max_{p_{k,n}} \ EE = \frac{\sum_{k=1}^{K}\sum_{n\in\Omega_k}\log_2(1+p_{k,n}H_{k,n})}{\tau\sum_{k=1}^{K}\sum_{n\in\Omega_k}p_{k,n}+P_c} \tag{5a}$$

$$\text{s. t. } p_{k,n} \geq 0 \ \forall k \in \mathrm{K}, n \in \mathrm{N} \tag{5b}$$

$$\sum_{k=1}^{K} p_{k,n} \leq P_T \tag{5c}$$

$$\Omega_i \cap \Omega_j = 0 \ \forall j \neq i \tag{5d}$$

$$\Omega_1 \cup \Omega_2 \cup \ldots \cup \Omega_K \subseteq \{1,2,\ldots,N\} \tag{5e}$$

$$\alpha_m R_1 - \alpha_1 R_m = 0 \ \forall m = 2,3,\ldots,K \tag{5f}$$

The problem in (5) belongs to a nonlinear fractional program. Fortunately, in the objective function (5a), the numerator is positive and concave function of $p_{k,n}$ while the denominator is positive and affine in $p_{k,n}$. These properties assert that the problem in (5) is a concave fractional program which can be transformed into a concave program by using Charnes-Cooper Transformation (CCT).

CCT Theorem [13]: A concave fractional program with

$$\max_{\mathbf{A}} \left\{ \frac{\Phi(\mathbf{A})}{\Psi(\mathbf{A})} \middle| \begin{matrix} \Omega(\mathbf{A}) \geq 0, \\ \Theta(\mathbf{A}) = 0 \end{matrix} \right\}, \tag{6a}$$

where $\Phi(\mathbf{A})$ and $\Omega(\mathbf{A})$ are concave functions while $\Psi(\mathbf{A})$and $\Theta(\mathbf{A})$ are affine functions. The above problem can be transformed to an equivalent concave program with

$$\max_{\boldsymbol{\beta},t} \left\{ t\Phi\left(\frac{\boldsymbol{\beta}}{t}\right) \middle| \begin{matrix} t\Omega(\boldsymbol{\beta}/t) \geq 0, t > 0, \\ \Theta(\mathbf{A}) = 0, t\Psi(\boldsymbol{\beta}/t) = 1 \end{matrix} \right\} \tag{6b}$$

under the transformation $t = 1/\Psi(\mathbf{A})$ and $\mathbf{A} = \boldsymbol{\beta}/t$. $\mathbf{A}$ and $\boldsymbol{\beta}$ are matrix-valued optimization variables while t is a scalar-valued parameter.

For ease of analysis, we define $N_k = |\Omega_k|$ as the total number of subcarriers assigned to user k and reorder subcarrier index such that subcarrier 1 through subcarrier n_1 are occupied by user 1, subcarrier $n_1 + 1$ through n_2 are occupied by user 2 and so forth. For user K, subcarrier $n_{K-1} + 1$ through subcarrier n_k are used. With the sorted subcarrier set N = $\{1,2,\ldots,n_1,n_1+1,\ldots, n_2\ldots,n_{K-1} + 1, \ldots, n_K\}$, the corresponding power allocation set and CNR set are given by P = $\{p_1,p_2,\ldots,p_N\}$ and H = $\{H_1,H_2,\ldots,H_N\}$, respectively. Then, substituting $p_n = y_n / t$ for $\forall n$ and $t = 1/[\tau \sum_{n=1}^{N} p_n + P_c]$, into problem in (5) transform to

$$\max_{\mathbf{y},t} f(\mathbf{y}, t) = t \sum_{n=1}^{N} \log_2 \left(1 + \frac{H_n y_n}{t}\right) \quad (7a)$$

$$\text{s. t. } p_n \geq 0 \; \forall k \in \mathrm{K}, \forall n \in \mathrm{N} \quad (7b)$$

$$\sum_{n=1}^{N} y_n - tP_T \leq 0 \quad (7c)$$

$$\alpha_k \sum_{n=1}^{n_1} \log_2 \left(1 + \frac{H_n y_n}{t}\right) - \sum_{n=n_{k-1}+1}^{n_k} \log_2 \left(1 + \frac{H_n y_n}{t}\right) = 0 \; \forall k \in \mathrm{K} \setminus \{1\} \quad (7d)$$

$$\tau \sum_{n=1}^{N} y_n + P_c t - 1 = 0 \quad (7e)$$

The conventional method for solving the problem (7) is to use the Lagrange multipliers as in [10], and form the cost function which is differentiated and made equal to zero. This yields the following

$$\omega_k = p_n + 1/H_n \; \forall k \in \mathrm{K} \text{ and } \forall n \in \Omega_k \quad (8)$$

where ω_k is denoted as the water-filling level for user k and can be further expressed in terms of ω_1 as

$$\omega_k = \frac{\left(\prod_{n=1}^{n_1} H_n\right)^{\frac{\alpha_k}{N_k}} \omega_1^{\frac{N_1 \alpha_k}{N_k}}}{\left(\prod_{n=n_{k-1}+1}^{n_k} H_n\right)^{\frac{1}{N_k}}} \; \forall k \in \mathrm{K} \quad (9)$$

The constraints in (7c) implies that the optimal solution lies inside and on the power constraint planes, respectively. Different from the conventional EE maximization schemes such as [14], the proposed energy-efficient method does not necessarily utilize all available power. This is intuitive since the amount of transmission power and EE are reciprocally related. That is, power should be consumed as less as possible in order to save energy. Nevertheless, the proposed scheme is flexible enough such that BS can enter full-power mode by transmitting power at the maximum level. In other words, the BS can operate in two modes: optimal-power and full-power. In the first mode, the BS can transmit power at less or equal to maximum power in order to achieve the highest EE. In the second mode, the BS is forced to transmit power at the maximum level in order to provision higher QoS requirements, but comes at the cost of reduced EE.

3.2.1. Optimal-Power Mode

When the associated Lagrange multiplier vanishes, we obtain

$$F_{\text{optimal}}(\omega_1) = \frac{\sum_{k=1}^{K} \alpha_k \log_2\left(\left(\prod_{n=1}^{n_1} H_n\right) \omega_k{}^{N_1}\right)}{P_c + \tau\left(\sum_{k=1}^{K} N_k \omega_k - \sum_{n=1}^{N} 1/H_n\right)} - \frac{\sum_{k=1}^{K} \alpha_k}{\tau \ln 2 \sum_{k=1}^{K} \alpha_k \omega_k} = 0 \tag{10}$$

It is obvious that ω_1 becomes the only relevant parameter when solving the EE optimization problem. To obtain the optimal ω_1^*, (10) can be solved by using a root-finding method. Optimal power allocation is then determined by (8) and (9) for $\forall k \in \mathrm{K}$ and $\forall n \in \Omega_k$.

3.2.2. Full-Power Mode

We now turn to the second case where the total transmission power lies on the power constraint plane. Define P_T^k as the total power allocated to user k with N_k subcarriers. Thus we have $\sum_{k=1}^{K} P_T^k = P_T$, which is equivalent to

$$F_{\text{full}}(\omega_1) = N_1\omega_1 + N_2\omega_2 + \cdots + N_K\omega_K - \left(\sum_{n=1}^{N} \frac{1}{H_n} + P_T\right) = 0 \tag{11}$$

Again, with (9), ω_1 becomes the only relevant coefficient while solving (11). As in optimal-power mode, we adopt a root-searching method to solve (11).

An unpleasant property of $F_{\text{optimal}}(\omega_1)$ in (10) and $F_{\text{full}}(\omega_1)$ in (11) is that the attained optimum ω_1^* does not take into account the adherence to positive power levels at this stage. One straightforward solution is to force $p_n = [\omega_k - 1/H_n]^+$ where $[x]^+ \triangleq \max\{x, 0\}$. However, when some subcarriers especially those with poor channel conditions, are assigned with negative power levels in the first place, the gap of power levels after applying aforementioned method will lead to the violation of rate proportionality constraints. According to water-filling theory [15], the water level should be recalculated when some subcarriers are assigned with negative power levels. More specifically, those subcarriers which experience deep fade will be omitted during the water-level recalculation. Recognizing this fact, the proposed method continues to drop the weakest subcarriers and recalculate ω_1 until there are no more assigned negative power levels. The complete steps of the proposed algorithm are described in *Algorithm 1*.

4. NUMERICAL RESULTS

In this section, we present simulation results to verify the effectiveness of the proposed scheme. Unless specified otherwise, the system settings are the same as in Table 1. All presented results below are an average of 1000 simulation runs, corresponding to 1000 different OFDM symbols. Here, each channel tap is modeled by an independent complex Gaussian random variable with its average power that follows the exponential power delay profile, while taking the time index of each channel tap by the integer multiples of sampling periods.

Algorithm 1. Proposed Energy-Efficient Resource Allocation

01: Adopt the subchannel allocation algorithm in 3.1 to distribute subcarriers among users Ω_k
02: Set $\Omega_k^{\text{optimal}} = \Omega_k$ for $\forall k \in \mathrm{K}$ and enter optimal-power mode
03: Update $\Omega_k^{\text{optimal}}$ for $\forall k \in \mathrm{K}$
04: Solve (10) to obtain ω_1^*
05: Compute $p_{k,n}^*$ with (8) and (9) for $\forall k \in \mathrm{K}\ \forall n \in \mathrm{N}$
06: if $p_{k,n}^* < 0$ for $\forall k \in \mathrm{K}\ \forall n \in \mathrm{N}$
07: drop the weakest subcarrier and go to line 03
08: end if
09: if $\sum_{k=1}^{K}\sum_{n=1}^{N} p_{k,n}^*$ for $\forall k \in \mathrm{K}\ \forall n \in \mathrm{N}$, $p_{k,n}^*$ is the solution.
10: else set $\Omega_k^{\text{full}} = \Omega_k$ for $\forall k \in \mathrm{K}$ and enter full-power mode
11: Update Ω_k^{full} for $\forall k \in \mathrm{K}$
12: Solve (11) to obtain ω_1^*
13: Compute $p_{k,n}^*$ with (8) and (9) for $\forall k \in \mathrm{K}\ \forall n \in \mathrm{N}$
14: if $p_{k,n}^* < 0$ for $\forall k \in \mathrm{K}\ \forall n \in \mathrm{N}$
15: drop the weakest subcarrier and go to line 11
16: end if
17: $p_{k,n}^*$ is the solution
18: end if

Table 1. OFDMA system settings

Parameters	Value
Total bandwidth	1 MHz
Subcarrier bandwidth B_n	15625 Hz
Sampling period	1 μs
Channel model	Multi-tap Rayleigh frequency-selective channel
Number of channel taps in time domain	6
Channel model of each tap	Clarke's flat fading model
Number of channel realizations for each tap	100
Power delay profile	[0 -8.69 -17.37 -26.06 -34.74 -43.43] dB
Relative delay	[0 1 2 3 4 5] μs
Number of subcarriers N	64
OFDM symbol duration	64 μs
Maximum Doppler frequency	100 Hz
Number of transmitted OFDM symbol	1000
Log-normal shadowing path loss	IEEE802.16d with Type C: flat terrain with light tree densities
Height of BS transmit antenna	30 m
Height of user receive antenna	2 m
Carrier frequency	2.9 GHz
Inverse of power amplifier efficiency τ	2.63
Circuit power P_C	10 W
Root-finding method	Bisection

4.1. Optimal-Power Mode vs. Full-Power Mode

Figure 1 shows the system performance versus available power. All six users are located with distances of 500 m, 550 m, 600 m, 650 m, 700 m and 750 m from the base station, respectively. N_0 is set to 10^{-8} W/Hz. By setting α_1:α_2:…:α_6 = 1:1:…:1, all users would obtain the same rate. The rationale behind power setting is that the maximum LTE user equipment (UE) transmission power and the maximum LTE BS transmission power are 23 dBm and 46 dBm, respectively [16]. From Figure 1 (a), it can be observed that optimal-power mode always outperforms full-power mode in terms of EE performance. From 20 dBm to 30 dBm, both schemes share the same EE because the amount of available power is too limited. In this case, the optimal EE lies on the power constraint plane, which is the same as in full-power mode. In the abundance of transmission power starting from 35 dBm, the optimal-power mode chooses not to transmit power at its maximum level in order to promote energy savings. In other words, the optimal-power mode maintains its power levels starting from 35 dBm to 50 dBm. In contrast, full-power mode always fully utilizes the available power, which degrades the EE performance. Based on these results, we can design a flexible power management in communication networks. When network is congested, we can always invoke the full-power mode such that more user demands can be fulfilled. On the other hand, we can reduce as much power consumption as possible when the network is idle. Interestingly enough, the EE performance trend of full-power mode is found to agree well the one presented in [14]. The crucial difference is that we can manage the power consumption in two modes flexibly. From Figure 1 (b), it can be seen that the SE increases as the available BS power increases. This is reasonable as power is proportional to the capacity. However, once exceeding 30 dBm, the sum capacity of optimal-power mode remains constant as it transmits at the same level of power. In contrast, the full-power mode pursues higher total capacity by utilizing all the available power.

4.2. Comparison with the Method in [10]

Figure 2 compares the system performance achieved by the proposed method with the method in [10] at different N_0 of 10^{-8}, 10^{-7}, 10^{-6} and 10^{-5} W/Hz, respectively. The larger the N_0, the lower the CNR of all subcarriers. As in the previous six-user system, we set $\alpha_1:\alpha_2:\ldots:\alpha_6 = 1:1:\ldots:1$ and $P_T =$ 5 W. From Figure 2 (a), it can be observed that the proposed method attains higher EE than the method in [10]. The reason is that the method in [10] cannot find accurately the optimal power distribution at N_0 of 10^{-7}, 10^{-6} and 10^{-5} W/Hz where the CNR of all users becomes relatively low. Rather, it can only solve the problem of (5) effectively at N_0 of 10^{-8}. Figure 2 (b) offers more insights into the SE performance of both methods. It is clear that the proposed scheme satisfies the proportional rate constraints in the strictest sense, while the method in [10] does not.

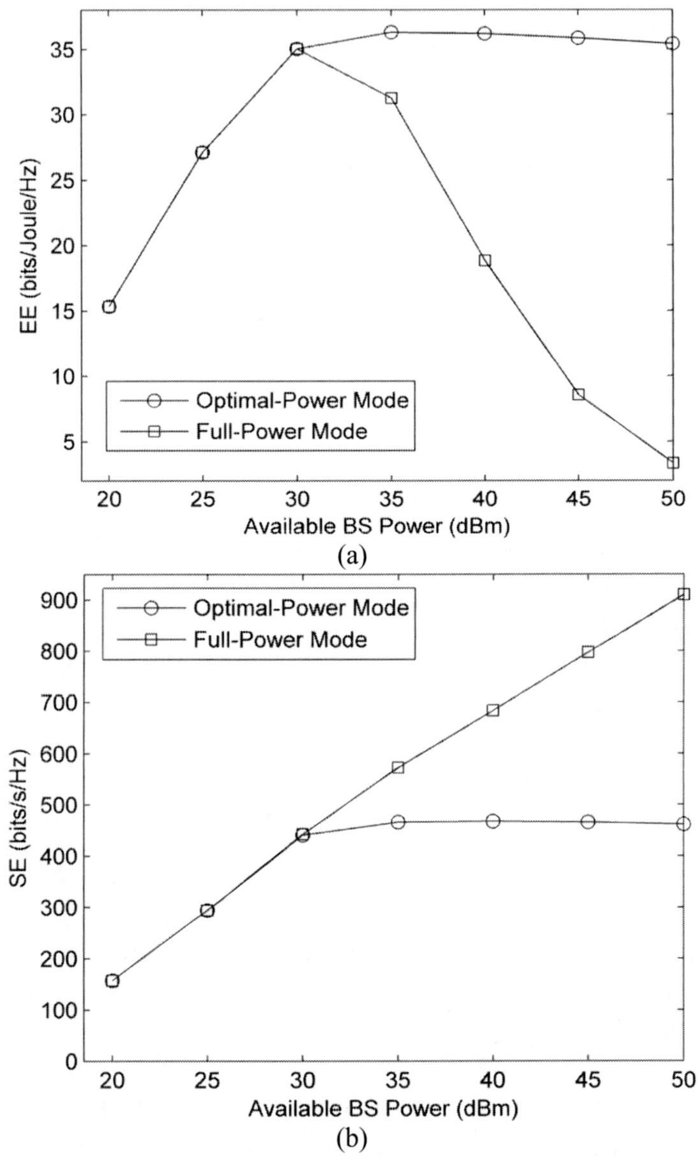

Figure 1. Network performance vs. available BS power for optimal-power mode and full-power mode. (a) Energy efficiency. (b) Spectral efficiency.

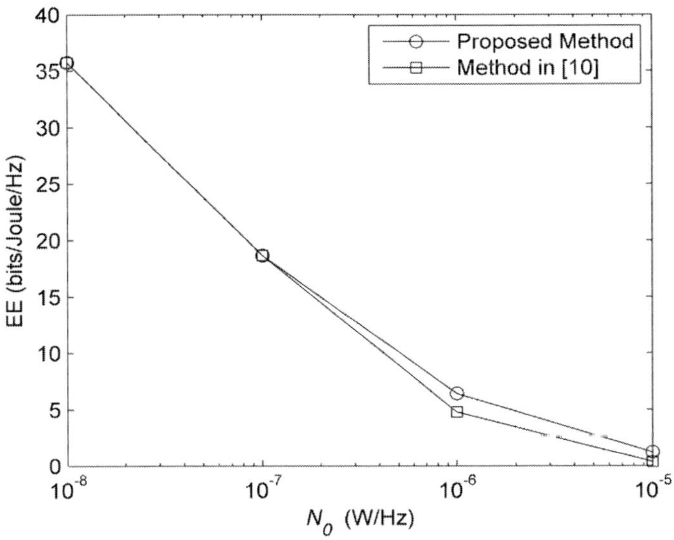

(a)

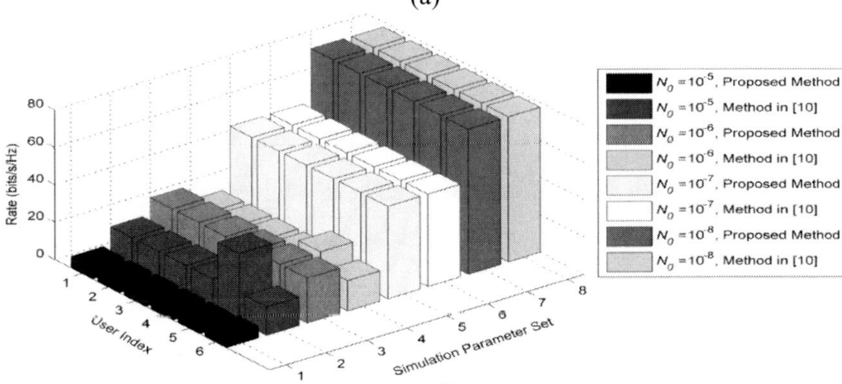

(b)

Figure 2. System performance vs. N_0. (a) EE (b) User data rate.

To measure how strictly rate constraints are satisfied, we first define $R_{k,i}$ and $\alpha_{k,i}$ as the data rate and proportionality rate constraint for user k at OFDM symbol i, respectively. Second, we let $\tilde{R}_{k,i} = R_{k,i}/\sum_{k=1}^{K} R_{k,i}$ be the normalized capacity for user k and $\tilde{\alpha}_{k,i} = \alpha_{k,i}/\sum_{k=1}^{K} \alpha_{k,i}$ be the normalized rate constraint for user k. Finally, the averaged normalized rate constraint deviation $\mathcal{D}$ is measured as [11]

$$\mathcal{D} = \sum_{i=1}^{I} \frac{\sum_{k=1}^{K} |\tilde{R}_{k,i} - \tilde{\alpha}_{k,i}|}{\max \sum_{k=1}^{K} |\tilde{R}_{k,i} - \tilde{\alpha}_{k,i}|} / I \qquad (12)$$

Table 2. Normalized rate constraint deviations for different number of users. $P_T = 5$ W and $N_0 = 1\times10^{-6}$ W/Hz

Number of users K	4	6	8	10
$\alpha_{1,i}:\alpha_{2,i}:\ldots:\alpha_{K,i}$	At OFDM symbol i, $\alpha_{k,i}$ is randomly drawn from the standard distribution in [1, 20]			
Proposed method	1.0509×10^{-16}	1.2001×10^{-16}	1.9419×10^{-16}	1.3512×10^{-16}
Method in [10]	0.5372	0.4284	0.2660	0.3549

Table 2 shows the normalized rate distributions for different number of users. In the settings, $I = 1000$ and $\alpha_{k,i}$ is drawn from the standard distribution in [1, 20] at OFDM symbol i. The rate constraint deviation of the proposed method is orders of magnitude much smaller than those achieved by the method in [10]. That is, the proposed method satisfies the rate proportionality constraint near the normalized rate constraint deviation of 10^{-16}. This implies that the method in [10] finds inaccurate solution in low subcarrier CNR. Considering that the proposed method can achieve energy-efficient communications regardless of arbitrarily channel characteristics, the proposed method is more useful in realistic environments.

CONCLUSION

An improved version of energy-efficient resource allocation scheme is proposed to maximize the energy efficiency of OFDMA systems under the constraints on total transmission power and rate proportionality. The proposed algorithm has the flavor of water-filling in which the algorithm loops until those subcarriers which are not energy efficient to carry any information bit are left out for transmission. In this way, the solution provides the hard guarantee of the proportional rate constraint without making any assumptions regarding the channel conditions. From the perspective of mobile operators, this feature is crucial as cell-edge subscribers typically have low received signal strength. Besides that, the strict satisfaction of user rate constraints ensures the full utilization of scarce radio resources. Furthermore, the proposed method offers flexibility to network operators in terms of energy saving modes. In the first mode, the BS consumes the least amount of transmission power. This can be useful when the user density is low. However, when the network is congested, the BS can enter full-power mode to increase transmission capacity.

REFERENCES

[1] Lorincz, J., Garma, T., and Petrovic, G. (2012). Measurements and Modelling of Base Station Power Consumption under Real Traffic Loads. *Sensors*, 12(12), 4281-4310. doi:10.3390/s120404281.

[2] Prabhu, R. S., and Daneshrad, B. (2010). An Energy-Efficient Water-Filling Algorithm for OFDM Systems. *Proceedings of the 2010 IEEE International Conference on Communications*, 1085-1089. doi: 10.1109/ ICC.2010.5502818.

[3] Isheden, C., and Fettweis, G. P. (2010). Energy-efficient multi-carrier link adaptation with sum rate-dependent circuit power. *Proceedings of the 2010 Global Telecommunications Conference*, 1-6. doi: 10.1109/GL OCOM.2010.5683700.

[4] Miao, G., Himayat, N., and Li, G. Y. (2010). Energy-efficient link adaptation in frequency-selective channels. *IEEE Transactions on Communications*, 58(2), 545-554. doi: 10.1109/TCOMM.2010.02.08 058 7.

[5] Illanko, K., Naeem, M., Anpalagan, A., and Androutsos, D. (2014). Frequency and Power Allocation for Energy Efficient OFDMA Systems with Proportional Rate Constraints. *IEEE Wireless Communications Letter*, 3(3), 313-316. doi: 10.1109/WCL.2014.032814.130888.

[6] Luo, F., Ye, Yu, Wu, Bin, Ho, P. H., and Ling, X. (2012). Power allocation based on fast Water-Filling for energy efficient OFDM and MIMO transmissions. *Proceedings of the 2012 IEEE Global Communications Conference*, 4758-4763. doi: 10.1109/GLOCOM.2012. 6503871.

[7] Xiao, X., Tao, X., and Lu, J. (2013). QoS-Aware Energy-Efficient Radio Resource Scheduling in Multi-User OFDMA Systems. *IEEE Wireless Communications Letter*, 17(1), 75-78. doi: 10.1109/WCL.2014.032814.1 30888.

[8] Zarakovitis, C. C., Ni, Q. (2016). Maximizing Energy Efficiency in Multiuser Multicarrier Broadband Wireless Systems: Convex Relaxation

and Global Optimization Techniques. *IEEE Transactions on Vehicular Technology*, 65(7), 5275-5286. doi: 10.1109/TVT.2015.2455536.

[9] Ren, Z., Chen, S., Hu, B., and Ma, W. "Energy-Efficient Resource Allocation in Downlink OFDM Wireless Systems With Proportional Rate Constraints," *IEEE Transactions on Vehicular Technology*, 63(5), 2139-2150. doi: 10.1109/TVT.2014.2311235.

[10] Chien, S. F., Tham, M.-L., Ting, T. O., and Kim, K. S. (2015). Energy-aware subchannel assignment in OFDMA system with proportional rate constraints. *Proceedings of the 24th Wireless and Optical Communication Conference*, 149-153. doi: 10.1109/WOCC.2015.73461 95.

[11] Shen, Z., Andrews, J. G., and Evans, B. L. (2005). Adaptive resource allocation in multiuser OFDM systems with proportional rate constraints. *IEEE Transactions on Wireless Communications*, 4(6), 2726-2737. doi: 10.1109/TWC.2005.858010.

[12] 3GPP TS 23.203 v. 10.8.0 (2012) *Policy and Charging Control Architecture*. http://www.etsi.org/deliver/etsi_ts/123200_123299/123203/10.08.00_60/ts_123203v100800p.pdf.

[13] Charnes, A., and Cooper, W. W. (1962). "Programming with Linear Fractional Functionals," *Naval Research Logistics Quarterly* 9.

[14] He, C., Li, G. Y., Sheng, B., and You, X. (2013). Energy and spectral efficiency of distributed antenna systems. *Proceedings of the* IEEE *Wireless Communications and Networking Conference*, 3225-3229. doi: 10.1109/WCNC.2013.6555079.

[15] Cover, T. M., and Thomas, J. A. (1991). *Elements of Information Theory*: John Wiley and Sons, Inc.

[16] 3GPP TR 36.931 v. 11.0.0 (2012) *Radio Frequency (RF) requirements for LTE Pico Node B*. http://www.qtc.jp/3GPP/Specs/36931-b00.pdf.

[17] Sadr, S., Anpalagan, A., and Raahemifar, K. (2009). Radio Resource Allocation Algorithms for the Downlink of Multiuser OFDM Communication Systems. *IEEE Communications Surveys and Tutorials*, 11(3), 92-106. doi: Doi 10.1109/Surv.2009.090307.

[18] Hasan, Z., Boostanimehr, H., and Bhargava, V. K. (2011). Green Cellular Networks: A Survey, Some Research Issues and Challenges. *IEEE Communications Surveys and Tutorials*, 13(4), 524-540. doi: 10.1109/SURV.2011.092311.00031.

[19] Fettweis, G. P., Zimmermann, E. (2008). ICT energy consumption-trends and challenges. *Proceedings of the 11th International Symposium on Wireless Personal Multimedia Communications*.

Biographical Sketches

Mau-Luen Tham, PhD

Department of Electrical and Electronic Engineering,
Lee Kong Chian Faculty of Engineering and Science,
Universiti Tunku Abdul Rahman (UTAR) Kajang, Malaysia

Research and Professional Experience: Radio Resource Allocation, Multimedia Communications, Network Optimization

Professional Appointments: Assistant Professor

Publications for the Last 3 Years:

Journals

1) Y.-H. Xu, W.-L. Wu, Y. Xu, M.-L. Tham, and N. Ramli, "A Framework of Fuzzy Control-Based Intelligent Control System for Greenhouse," *Artificial Intelligence Research*, vol. 6, no. 1, pp. 1-5, 2017.
2) M.-L. Tham, C.-O. Chow, Y.-H. Xu, and N. Ramli, "Two-Level Scheduling for Video Transmission over Downlink OFDMA Networks," *PLOS ONE*, vol. 11, no. 2, pp. 1-15, Feb. 2016.
3) M.-L. Tham, C.-O. Chow, and Y.-H. Xu, "Impact of Random Packet Arrival on the Performance of Video Streaming over OFDMA Downlink Systems," *Wireless Pers Commun*, vol. 85, no. 3, pp. 655-669, Dec. 2015.
4) M.-L. Tham, C.-O. Chow, Y.-H. Xu, K. N. Choong, and C. S. Lee, "Seamless handover between unicast and multicast multimedia streams," *J. Zhejiang UnivSci. C*, vol. 15, no. 10, pp. 929-942, Oct. 2014.
5) Y.-H. Xu, C.-O. Chow, G.-X. Kok, and M.-L. Tham, "A Load-aware Scheme for Providing Heterogeneous Multimedia Broadcast/Multicast Service (Het-MBMS)," *Wireless PersCommun.*, vol. 77, no. 3, pp. 2391-2414, Aug. 2014.
6) M.-L. Tham, C.-O. Chow, M. Iwahashi, and H. Ishii, "BER-driven resource allocation for scalable bitstreams over OFDMA networks," *IEEE Trans. Veh. Technol.*, vol. 63, no. 6, pp. 2755-2768, July 2014.

7) Y.-H. Xu, C.-O. Chow, M.-L. Tham, and H. Ishii, "An Enhanced Framework for Providing Multimedia Broadcast/Multicast Service (MBMS) over Heterogeneous Networks," *J. Zhejiang UnivSci. C*, vol. 15, no, 1, pp. 63-80, Jan. 2014.

Conferences

1) K. N. Choong, D. Naghshbandi, M.-L. Tham, and H. H. M. Shariff, "Wireless Presentation Management System," In: *Proc. The Fourth International Conference on E-Learning and E-Technologies in Education (ICEEE2015)*, Indonesia, Sep. 2015.
2) S. F. Chien, M.-L. Tham, T. O. Ting, and K. S. Kim, "Energy-Aware Subchannel Assignment in OFDMA Systems with Proportional Rate Constraints," *24th Wireless and Optical Communication Conference (WOCC)*, Taiwan, June 2015.
3) W.-K. New, C.-O. Chow, M.-L. Tham, and G.-X. Kok, "Robust framework for disaster mitigation using heterogeneous networks," *MJIIT-JUC Joint International Symposium*, Kuala Lumpur, 12-13 Nov. 2014.
4) M.-L. Tham, C.-O. Chow, K. Utsu, and H. Ishii, "BER-driven resource allocation in OFDMA systems," In: *Proc. IEEE 24th Personal Indoor and Mobile Radio Communications (PIMRC)*, London, 8-11 Sept. 2013.

Su Fong Chien, PhD

Department of Strategic Advanced Research (StAR),
MIMOS Berhad, Kuala Lumpur, Malaysia

Research and Professional Experience: Radio Resource Allocation, Network Optimization

Professional Appointments: Principal Researcher

Publications for the Last 3 Years:

Journals

1) S. F. Chien, T. O. Ting, X.-S. Yang, K. Takahashi, "Mathematical analysis of energy efficiency optimality in multi-user OFDM

systems," *Wireless Communications and Mobile Computing*, vol. 16, no. 3, pp. 252-263, 2016.
2) T. O. Ting, S. F. Chien, X.-S. Yang, S. Lee, "Analysis of quality-of-service aware orthogonal frequency division multiple access system considering energy efficiency," *IET Communications*, vol. 8, no. 11, pp. 1947-1954, 2014.

Conferences

1) S. F. Chien, M.-L. Tham, T. O. Ting, and K. S. Kim, "Energy-Aware Subchannel Assignment in OFDMA Systems with Proportional Rate Constraints," *24th Wireless and Optical Communication Conference (WOCC)*, Taiwan, June 2015.
2) S. F. Chien, K. K. Lo, K. H. Kwong, H. S. Lim, "Performance Evaluation for Sum Capacity in OFDMA Systems with Proportional Fairness," *14th IEEE International Conference on Computer and Information Technology (CIT)*, Xi'an, China, 2014.
3) S. F. Chien, T. O. Ting, X.-S. Yang, A. K. N. Ting, D. W. Holtby, "Optimizing energy efficiency in multi-user OFDMA systems with genetic algorithm," *International Conference on Advances in Computing, Communications and Informatics (ICACCI)*, India, 2013.

In: Energy Efficiency
Editor: Julian E. Perez
ISBN: 978-1-53611-040-1

Chapter 4

THE DESIGN AND OPTIMIZATION FOR A CLOSED-LOOP PRESSURE RETARDED OSMOSIS PROCESS

Ali Altaee[1], John Zhou[1], Adnan Alhathal AlAnezi[2] and Guoyong Zhao[3]

[1]School of Civil and Environmental Engineering,
University of Technology Sydney, Sydney, Australia
[2]Department of Chemical Engineering Technology,
College of Technological Studies,
The Public Authority for Applied Education
and Training (PAAET), Kuwait
[3]Department of Mechanical Engineering,
Shandong University of Technology, Zibo, Shandong, China

ABSTRACT

Pressure Retarded Osmosis (PRO) is an emerging renewable energy technique with potential for scale up and commercialization. It has been extensively investigated in the last decade as a new technique for power generation from osmotic energy. Both open-loop and closed-loop PRO have been suggested for power generation, differing in that the draw and feed solutions are disposed of after membrane treatment in the open-loop PRO process. The closed-loop PRO (CLPRO) process operates like a heating engine in which the diluted draw solution enters a special unit after leaving the membrane for regeneration and reuse. CLPRO uses a pure draw solution for process operation, hence reducing membrane fouling and cleaning requirements. Furthermore, there is no waste discharge to the environment and no pretreatment is required for the feed solution.

CLPRO design requires optimizing the operating parameters to reduce the capital and operating costs of the process. Membrane area, feed pressure, draw solution, and draw solution flow rate are among the key operating parameters that significantly impact CLPRO performance. A computer model is developed in the current study to evaluate the impact of these parameters on CLPRO performance. The feed solution osmotic pressure was considered to be negligible because of the high purity of water produced from the regeneration unit. Simulation results revealed that feed pressure is the most influential parameter on the CLPRO performance, followed by the draw solution flow rate and membrane area. In fact, the membrane impact on the performance of CLPRO did not exceed 4%. The result of this study should be considered in the early stages of CLPRO design to reduce operating and capital costs.

Keywords: pressure retarded osmosis, closed-loop pressure retarded osmosis, salinity gradient resource, osmotic power plant, renewable energy

INTRODUCTION

There are increasing demands on energy supplies as they become an essential requirement to support modern life in most countries. Conventional resources for power generation that depend on fossil fuels become less reliable because of price fluctuations and long-term supply uncertainties [1-5]. Furthermore, strong environmental awareness coupled with recent efforts to reduce greenhouse gas emissions has encouraged a shift from fossil fuels to renewable power generating resources. This has resulted in a significant investment in renewable energy technologies such as wind, solar, hydro and fuel cells [6-9]. Power generation from renewable resources is constantly increasing to compensate for conventional fossil fuel technologies. This has encouraged researchers and scientists to seek new competitive resources of renewable energy that have scale up and commercialization potential [10-11]. A promising renewable technology that has been extensively investigated in the last decade is pressure retarded osmosis (PRO) [12-15]. Many studies have been published evaluating its technological performance and competitiveness with existing renewable energy technologies and future applications for power generation [16-19]. PRO has many advantages over conventional wind and solar renewable energy technologies, such as not being affected by solar radiation or wind speed, can operate 24 hours per day, a compact system, and low foot print.

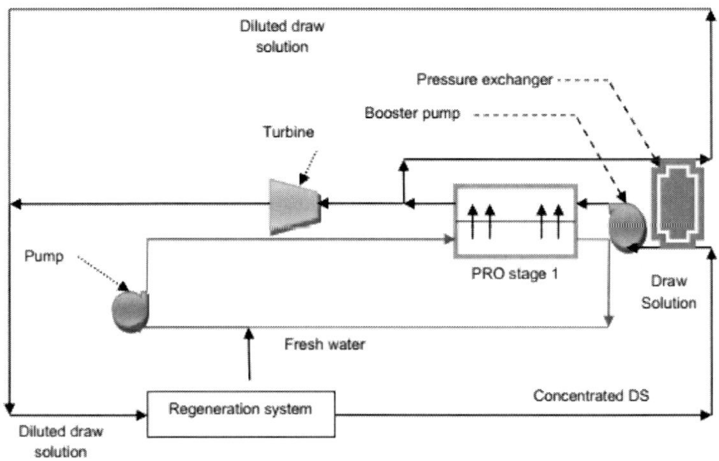

Figure 1. Close-loop PRO process diagram.

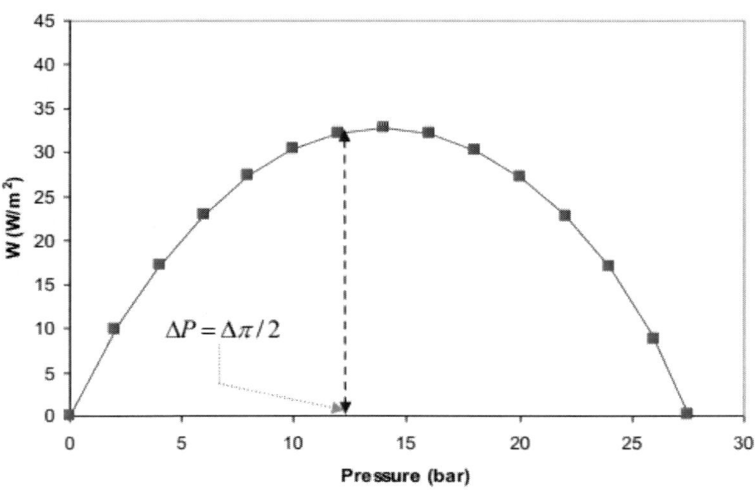

Figure 2. Relationship between feed pressure and PRO power density. Maximum power density (W) is achieved at $\Delta P = \Delta\pi / 2$. Feed solution is DI water, draw solution is 43 g/L NaCl, and the applied feed pressure is 14 bar.

PRO relies on the osmotic energy of a salinity gradient resource for power generation; it converts the chemical potential into hydraulic pressure as fresh water transports from the feed to the pressurized draw solution. In the PRO process, two solutions of different concentrations and osmotic pressures are separated by a semi-permeable membrane with a high rejection rate and water flux [15]. The high concentration solution is usually named the draw solution and the low concentration solution is called the feed solution. The draw solution is slightly pressurized before entering the PRO module. Inside the PRO membrane, water transports from the feed to the draw solution side due to the osmotic pressure gradient across the membrane. As such the draw solution gets more diluted as it leaves the PRO membrane. For an economically feasible PRO process, part of the pressurized draw solution flow returns to a pressure exchanger after leaving the membrane to pressurize the raw draw solution for the PRO process (Figure 1). The rest of the pressurized draw solution goes to a hydro turbine system for power generation. Different scenarios have been proposed for salinity gradient resource management at the end of the PRO process [20-21]. In the case of open-loop PRO (OLPRO), the draw and feed solutions are discharged to the marine environment [12]. For the closed-loop PRO process (CLPRO), the draw solution goes to a regeneration unit for recycling and reuse [12, 22]. The system operates like a heat engine with potential for applications especially when there is a valid source of waste heat.

In principle, the PRO draw solution has to be pressurized before entering the membrane module. The applied pressure on the draw solution side is typically lower than the osmotic pressure difference across the membrane. One of the key factors in determining the optimum draw solution hydraulic pressure is the membrane power density, W (W/m^2). The latter is a theoretical value representing the amount of power generated per square meter of the membrane. Experimental work at the laboratory scale has demonstrated that power density, W, reaches a maximum when the hydraulic pressure is equal to half of the osmotic pressure gradient across the PRO membrane ($\Delta\pi/2$). In other words, W reaches a maximum when $\Delta P = \Delta\pi/2$ as shown in Figure 2. Technically, the higher the power density of the PRO process, the higher the process efficiency. A pilot plant study carried out by Statkraft proposed that power density should equal or exceed a theoretical value of 5 W/m^2 for the PRO process to be economically feasible [23-24]. However, this theoretical value was based on the Statkraft experimental pilot plant that used fresh water and seawater from fjords as a salinity gradient resource. New developments in PRO membrane technology and our understanding of the performance of PRO technology may question the reliability of their estimate. However, it should be mentioned that the major breakthrough in the Statkraft pilot plant was the inclusion of a pressure exchanger (PX) to exchange the hydraulic pressure of the diluted draw solution with the raw draw solution (Figure 1). The anticipated advantage of this design was to reduce the PRO operation cost. Unfortunately, the Statkraft PRO plant shut down in 2011; Statkraft did not reveal the reasons, but it is thought to be due to the membrane flux caused by an insufficient osmotic pressure gradient across the membrane.

Subsequent works by Saito et al. [21] demonstrated that a high performance PRO process was achieved using a proper PRO membrane and salinity gradient process. The PRO membrane was provided by Toyobo Membrane Company, and reverse osmosis (RO) brine was coupled with tertiary treated wastewater as the salinity gradient resource. A three port hollow fiber membrane was initially used in the pilot plant but was replaced with a four port hollow fiber membrane because of its greater resistance to fouling [21]. The mega-ton PRO power plant was a great success; a 13.5 w/m^2 power density was achieved in the pilot plant tests, which is two and a half times more than proposed by the Statkraft Company. The applied hydraulic pressure of the draw solution was 25 bar, which indicated the high mechanical strength of the PRO membrane. The net power output of the plant was 100 kW and the running cost was reasonably inexpensive at 0.061$/kWh [25-26]. The mega-ton pilot plant has demonstrated that a 5 W/m^2 power density can be easily achieved using a proper PRO membrane and salinity gradient resource.

The CLPRO process has also been proposed for power generation from salinity gradient resources [12]. The expected advantages of CLPRO are draw solution recycling and reuse, no solution discharge to the environment, no need for pretreatment, and benefits from any source of waste heat. In addition, CLPRO performs better than OLPRO because of the negligible concentrative concentration polarization on the feed solution side [27]. This is mainly due to the high purity of feed solution, which is normally the fresh/deionized water from the regeneration unit. The CLPRO system draw solution of CLPRO system could be an inorganic metal salt such as sodium chloride, magnesium sulfate, or potassium chloride. Thermolytic draw solutions such as ammonium carbon dioxide have also been proposed for CLPRO, especially when there is a source of waste heat [28]. Ammonium carbon dioxide, for example, can be regenerated at 70°C; this makes thermal evaporators a perfect technique for draw solution regeneration and reuse. In general, thermolytic draw solutions, such as ammonium based solutions, exhibit a reported power density of 13.5 W/m^2, which is two and a half times more than that proposed by the Statkraft Company's moderate osmotic pressure with inorganic draw solutions such as sodium chloride and magnesium chloride. However, it should be mentioned that no pilot plant work on CLPRO has yet been carried out.

The current study investigates the optimization of the CLRPO process, taking into account the impact of key performance parameters such as the membrane area, draw solution flow rate, draw solution concentration and hydraulic feed pressure. The effect of each parameter was individually investigated using a computer model. The results from this study should be considered in the design of a CLPRO system to reduce the capital and operating costs. These results are also crucial to compare the performance of PRO processes under different operating conditions.

Process Modeling

Natural osmosis is the main driving force for solvent transport across the membrane in the PRO process. Typically, water transports from the low to the high concentration solution; these are the feed and draw solutions. Solution diffusion theory has been applied for estimating the membrane flux in the PRO process:

$$J_w = A_w(\Delta\pi - \Delta P) \tag{1}$$

where J_w is the membrane flux (L/m^2h), A_w is the water permeability coefficient (L/m^2hbar), $\Delta\pi$ is the osmotic pressure difference across the membrane (bar), and ΔP is the hydraulic pressure (bar). Equation 1 is valid for calculating water flux in an ideal PRO membrane. Previous studies, however, have shown that experimental water flux is much lower than the theoretical water flux estimated from equation 1 because of the adverse impact of internal and external concentration polarization. Equation 2 has been applied to predict permeate flux in the PRO membrane taking into account the effects of internal and external concentration polarizations [16, 29]:

$$Q_P = A * A_w\left(\frac{\pi_{Db} * e^{\frac{-(Q_P/A)}{k}} - \pi_{Fb} * e^{\frac{Q_P}{A}K}}{1 + \frac{B}{\frac{Q_P}{A}}\left(e^{\frac{Q_P}{A}K} - e^{\frac{-(Q_P/A)}{k}}\right)} - \Delta P\right) \quad (2)$$

where Q_P is the permeate flow rate (m^3/h), A is the membrane area (m^2), π_{Db} and π_{Fb} are the osmotic pressures of the bulk draw and feed solutions (bar), respectively, A_w is the water permeability coefficient, ΔP is the hydraulic pressure across the PRO membrane (bar), k is the mass transfer coefficient (m/s), B is the solute permeability coefficient (m/h), and K is the solute resistivity for diffusion within the porous support layer (s/m). The membrane flux, J_w (L/m^2h), was then calculated as the ratio of permeate flow to the membrane area; i.e., $J_w = Q_p/A$. The feed solution in the CLPRO process is distilled water; this suggests that $\pi_{Fb} \ll \pi_{Db}$ and was thus assumed to be zero. As illustrated in Figure 2, the optimum PRO performance is achieved at $\Delta P = \Delta\pi / 2$; $\Delta\pi$ is average osmotic pressure of the bulk feed and draw solutions. As such, π_{Db} was calculated from equation 2:

$$\pi_{Db} = 2 * \Delta P + \pi_{Fb} \quad (3)$$

In the current study, the Van't Hoff equation was applied to estimate the osmotic pressure of the draw solution as in the following equation 29:

$$\pi_{Db} = 1.12(273 + T)\Sigma m_n \quad (4)$$

where, T is feed temperature (Kelvin) and m_n is the molar concentration of the nth ion species. Inorganic metal salts have been proposed as draw solutions for the PRO process; NaCl was the draw solution of our PRO process because of its high osmotic pressure, low cost, availability, and the ease of estimating this parameter. Equation 4 can be re-arranged to calculate the concentration of the bulk draw solution in mg/L as in the following equations:

$$\pi_{Db} = \frac{C_{Nab} * 1.12 * t}{M_{wt\,Na} * 14.5} + \frac{C_{Clb} * 1.12 * t}{M_{wt\,Cl} * 14.5} \quad (5)$$

$$C_{Clb} = \frac{M_{wt\,Cl}}{M_{wt\,Na}} C_{Nab} = 1.54 C_{Nab} \tag{6}$$

$$C_{Nab} = \frac{\pi_{Db}}{\left(\frac{1.12T}{M_{wt\,Na} * 14.5}\right) + \left(\frac{1.12 * 1.54 * T}{M_{wt\,Cl} * 14.5}\right)} \tag{7}$$

where M_{wtNa} and M_{wtCl} are the molecular weights of *Na* and *Cl* ions, respectively (mg/M), and C_{Nab} and C_{Clb} are the bulk concentration of *Na* and *Cl* ions, respectively (mg/L). The concentration of the draw solution at the PRO membrane inlet was calculated from the mass balance equation:

Mass and flow balance equation on the other hand, the draw solution inlet concentration was calculated from the mass and flow balance equation around the draw solution side of the PRO membrane as follows:

$$C_{Di} * Q_{Di} + C_P * Q_P = C_{Do} * Q_{Do} \tag{8}$$

where C_P is the permeate concentration (mg/L). C_{Do} in equation 8 can be expressed in terms of bulk and inlet concentrations of draw solution as in the following equation:

$$C_{Do} = 2 * C_{Db} - C_{Di} \tag{9}$$

The bulk concentration of the draw solution is the average of the inlet and out concentrations of the draw solution:

$$C_{Db} = \frac{C_{Di} + C_{Do}}{2} \tag{10}$$

Practically, $C_{P1} << C_{Do1}$ and C_{Di1}, hence it can be ignored. Substituting equation 9 in equation 8:

$$C_{Di} * Q_{Di} = (2 * C_{Db} - C_{Di}) * Q_{Do} \tag{11}$$

$$C_{Di} = \frac{2 * C_{Db} * Q_{Do}}{Q_{Di} + Q_{Do}} \tag{12}$$

where C_{Do} and C_{Di} are the inlet and outlet draw solution concentrations (mg/L), respectively, Q_{Do} and Q_{Di} are the inlet and outlet draw solution flow rates (L/h), respectively, and C_P is the concentration of permeate flow (mg/L). In the current study, C_{Do} was estimated from substituting C_{Di} from equation 12 in equation 9. The power density of the PRO process, W (W/m^2), was calculated from the following expression:

$$W = J_w * \Delta P \quad (13)$$

Furthermore, power, P_w (kW), generated by the PRO process is calculated as a function of the permeate flow rate and the hydraulic pressure of the draw solution:

$$P_w = Q_P * \Delta P \quad (14)$$

It was envisaged that the PRO recovery rate, Re (%), should be calculated as an indicator of the net water transport due to the osmotic pressure gradient across the PRO membrane:

$$\mathrm{Re} = \frac{Q_P}{Q_F} \quad (15)$$

The draw solution concentration was determined at varied testing conditions: membrane areas, hydraulic pressures and draw solution flow rates. As a matter of fact, the draw solution osmotic pressure is the main driving force across the CLPRO process because the feed water is deionized water. Therefore, variations of the membrane area, hydraulic feed pressure and draw solution flow rate require an adjustment of the draw solution concentration to reduce operation cost.

Model Calibration

Model calibration was carried out using data from previous experimental work on a flat sheet frame and plate PRO unit [30]. Different concentrations of draw and feed solutions were tested and model and experimental power densities, W_{modl} and $W_{expt,}$ respectively, were evaluated. Practically, power density is one of the key parameters used in the evaluation of the PRO process performance.

Table 1 shows the testing parameters of the PRO process that were used in the model calibration. The membrane permeability coefficient was for a cellulose acetate PRO membrane made by Hydration Technology Innovation (HTI), USA. Table 2 compares the W_{modl} *and* W_{expt}; Table 1 shows good agreement between the model and experimental results, with less than a 10% difference between them. Interestingly, the difference between W_{modl} *and* W_{expt} for DI water feed solution remained approximately 7% despite a 50% increase of hydraulic pressure (from 10 bar to 15 bar). Using a DI water feed solution resembles CLPRO the situation, which uses DI water as a feed solution. In general, the model was considered reliable in estimating the performance of PRO processes.

PERFORMANCE OF THE CLPRO PROCESS

The PRO process performance is affected by a number of technical and operating parameters that significantly affect the energy output and operation cost. Membrane area, feed pressure and draw solution flow rate are some of the key parameters investigated in the present study.

Table 1. The testing parameters for the PRO model

Parameter	k (m/h)	K (h/m)	A_w (m/h.bar)	B (m/h)
Value	0.306	115-125	$6.7*10^{-4}$	$4*10^{-4}$

Table 2. Comparison between model and experimental power density

Draw Con. (g/L)	Feed Con. (g/L)	P (bar)	W_{expt} (W/m^2)	W_{modl} (W/m^2)	% Diff. W
35	0	10	2.8	2.6	7.1%
	2.5	12	2.3	2.1	8.7%
	5	11	1.7	1.55	8.8%
60	0	15	6.8	6.3	7.3%
	2.5	23	6.5	6.1	6.1%
	5	15	4.1	3.7	9.7%

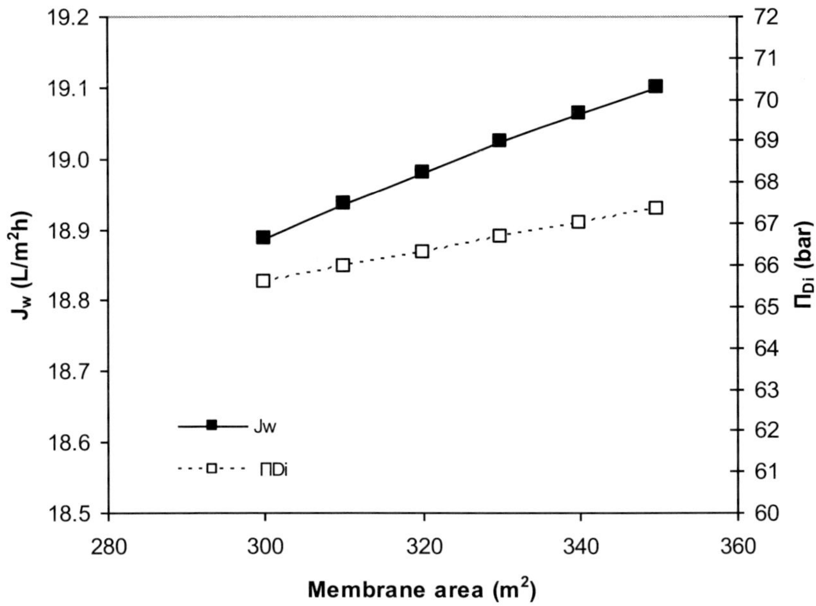

A

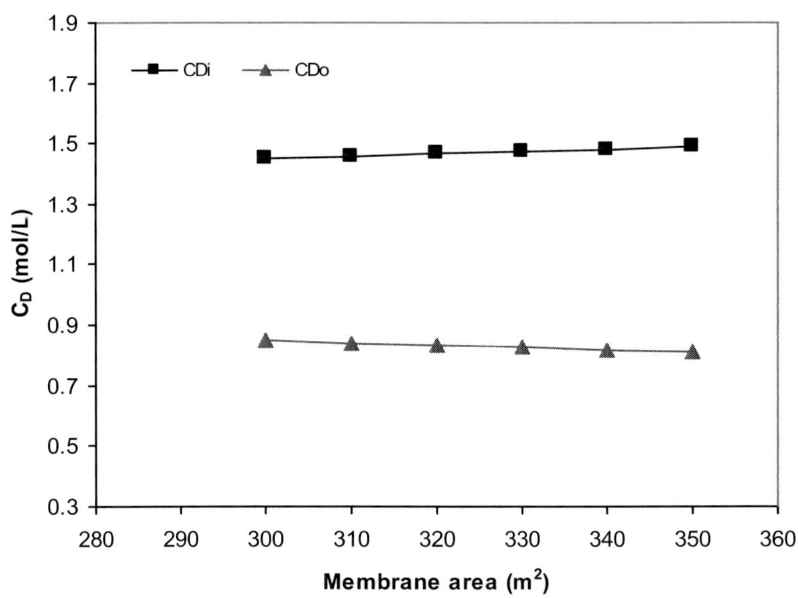

B

Figure 3. (Continued).

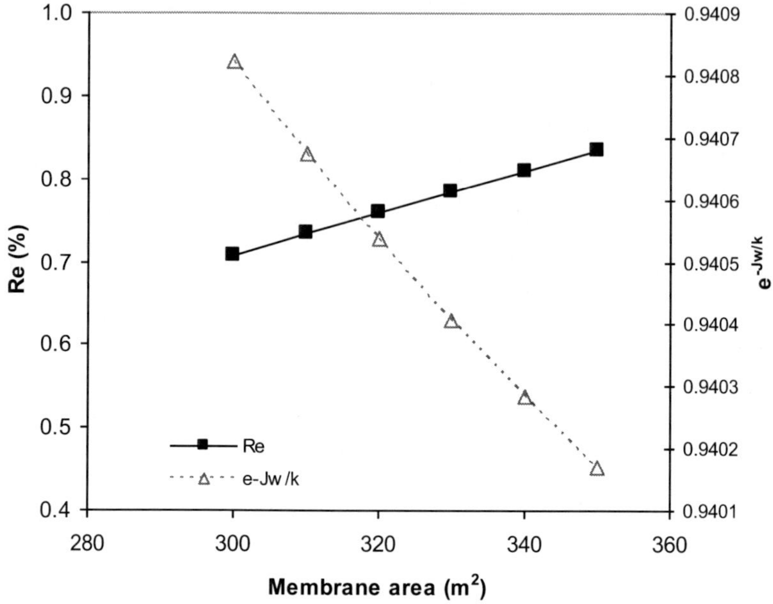

C

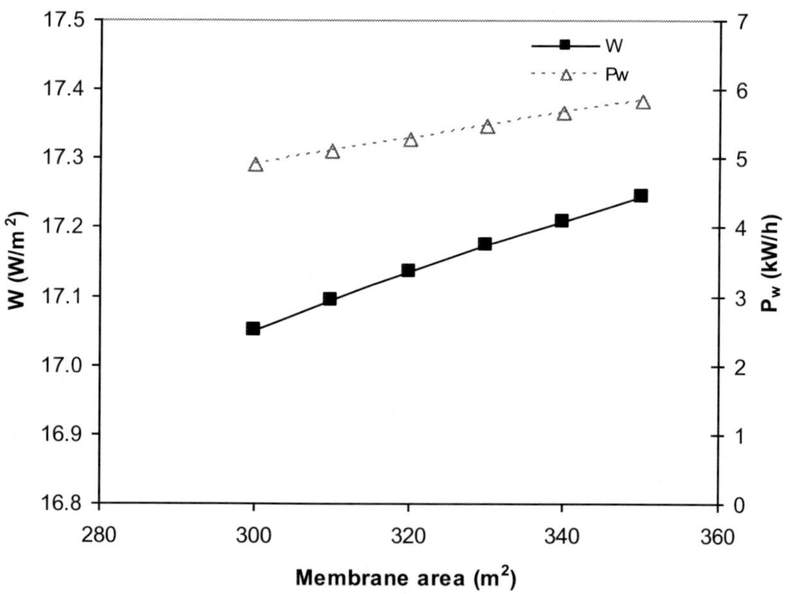

D

Figure 3. The impact of membrane area on the CLPRO performance A) membrane flux and osmotic pressure B) concentration of draw solution C) recovery rate and dilutive concentration polarization D) power density and power generation.

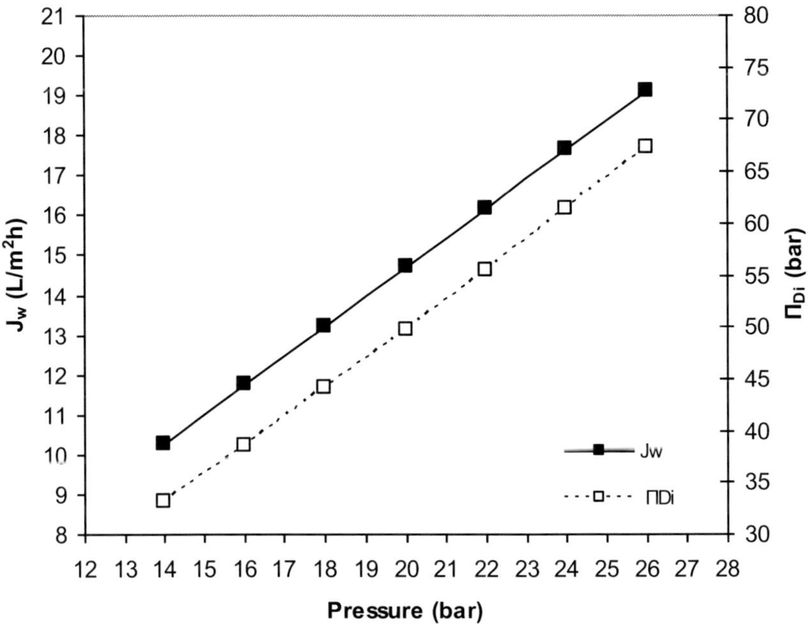

A

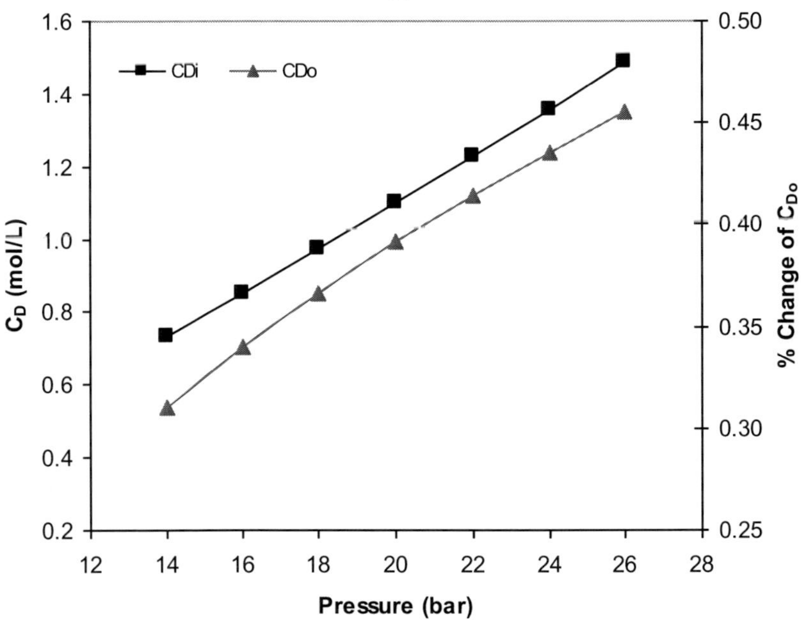

B

Figure 4. (Continued).

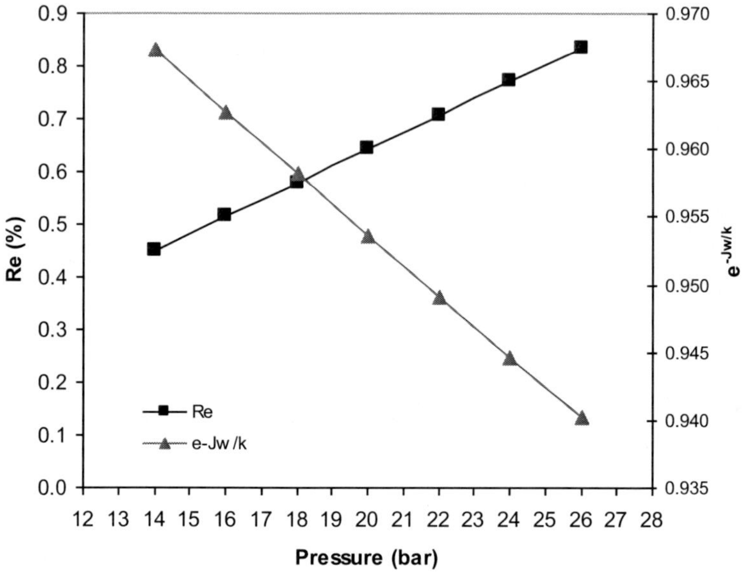

C

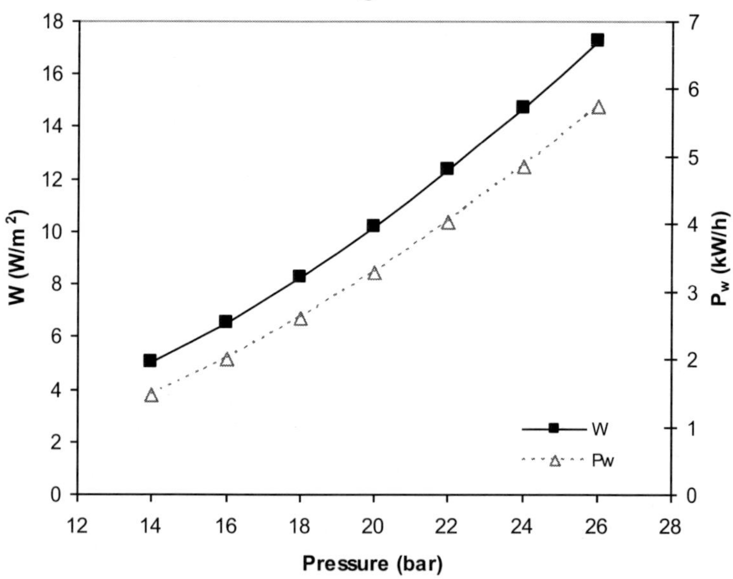

D

Figure 4. The impact of feed hydraulic pressure on the PRO performance A) membrane flux and osmotic pressure B) concentration of draw solution C) recovery rate and dilutive concentration polarization D) power density and power generation.

IMPACT OF MEMBRANE AREA

The impact of membrane area on the PRO performance was evaluated under the following conditions: 26 bar feed pressure and 8000 L/h draw solution flow rate. Membrane flux slightly increased with increasing membrane area, from 18.9 L/m^{2}h at 300 m^2 to 19.1 L/m^{2}h at 350 m^2 (Figure 3A). As predicted by the model, this increase corresponded to a slight concentration increase from 1.45 mol/L to 1.49 mol/L at 300 m^2 and 350 m^2, respectively (Figure 3B). This increase in the draw solution concentration resulted in a corresponding increase of the draw solution osmotic pressure. The osmotic pressure increased from 65 bar at 300 m^2 to 67 bar at 350 m^2 (Figure 3A). The increase of membrane flux caused a decrease in the concentration of the outlet draw solution, C_{Do}, from 0.85 mol/L to 0.81 mol/L at 300 m^2 and 350 m^2, respectively (Figure 3B). Decreasing the concentration of C_{Do} may affect the PRO operating cost depending on the regeneration technology of the draw solution. Interestingly, the membrane area increase resulted in a direct increase of the modulus of dilutive concentration polarization, $e^{-J_w/k}$ (Figure 3C). An $e^{-J_w/k}$ value close to unity is indicative of no or negligible dilution concentration polarization; $e^{-J_w/k}$ deviation from unity refers to the development of dilutive concentration polarization at the solution-membrane surface interface. Practically, water permeation from the feed to the draw solution side results in the dilution of draw solution concentration at the boundary layer, thus increasing the $e^{-J_w/k}$. The recovery rate of the PRO process, (%Re), increased with the increase of membrane area; %Re was 71% and 84% at 300 m^2 and 350 m^2, respectively (Figure 3C). %Re is calculated from equation 14; for a given feed flow rate (8000 L/h), the increase of %Re was induced by the higher membrane flux across the membrane (Figure 3A). It should be mentioned that the effect of concentrative concentration polarization was negligible in the CLPRO process because of the DI water feed solution.

Finally, the impact of membrane area on the power density, W, and power generation, P_w, of the PRO process is presented in Figure 3D. Increasing membrane area caused a slight increase of the power density; W was 17 W/m^2 at 300 m^2 and 17.2 W/m^2 at 350 m^2. In other words, the 16% increase of membrane area resulted in only a 1% increase of W. This suggests that the membrane area has a limited impact on W, which is an indicator of the performance of the PRO process. Additionally, P_w increased from 5.1 kW/h to 6 kW/h with the increase of membrane area from 300 m^2 to 350 m^2

(Figure 3D). These results support the conclusion that membrane area has minimum impact on the performance of the PRO process. For the rest of this study, 350 m^2 membrane area was applied because of its higher performance in the CLPRO process.

IMPACT OF HYDRAULIC PRESSURE

Hydraulic feed pressure is one of the key parameters that affect the performance of the PRO process [30, 31]. Figure 4 shows the impact of hydraulic pressure on the CLPRO performance at a 350 m^2 membrane area and 8000 L/h draw solution flow rate. There was a tangible increase in the membrane flux with increasing feed pressure (Figure 4A); membrane flux increased from 10.3 L/m^2h at 14 bar to 19.1 L/m^2h at 26 bar, an approximately 85% increase in the membrane flux. This higher membrane flux caused by increasing the inlet osmotic pressure from 33 bar at 14 bar to 67 bar at 26 bar (Figure 4A). The draw solution inlet concentration, C_{Di}, caused this increase of osmotic pressure: C_{Di} increased from 0.73 mol/L at 14 bar to 1.49 mol/L at 26 bar (Figure 4B). On the other hand, the outlet concentration of draw solution increased with increasing the draw solution hydraulic pressure from 14 bar to 26 bar (Figure 4B); this was due to the high membrane flux at 16 bar hydraulic pressure.

The recovery rate, %Re, of the PRO process was calculated and is presented in Figure 4C. %Re increased from 45% at 14 bar to 84% at 26 bar; this was an 87% increase of the PRO recovery rate and it is attributed to the high membrane flux. However, the dilutive concentration polarization, $e^{-J_w/k}$, decreased with the increase of hydraulic pressure, indicating more severe concentration polarization (Figure 4C). This result is expected because of the gradual increase of membrane flux with the increasing hydraulic pressure. In fact, the $e^{-J_w/k}$ value decreased from 0.97 at 14 bar hydraulic pressure to approximately 0.94 at 26 bar. The impact of hydraulic pressure on the power density of the PRO process is shown in Figure 4D. Power density, W, increases from 5 W/m^2 at 14 bar to approximately 17 W/m^2 at 26 bar; this was almost 3.5 times higher. The minimum W for an economically feasible PRO process is 5 W/m^2; as such, a 14 bar hydraulic pressure is probably the minimum operating hydraulic pressure to justify the PRO process. It should be noted that using a PRO membrane with a higher permeability coefficient would improve the performance and power density of the PRO process; i.e., W

may exceed 5 W/m^2 at lower hydraulic pressure for the same testing conditions in Figure 4. Finally, power generation, P_w, increased 3.5 times when increasing the hydraulic pressure from 14 to 26 bar (Figure 4D). P_w was 1.75 kW/h at 14 bar hydraulic pressure but increased to 6 kW/h at 26 bar.

IMPACT OF DRAW SOLUTION FLOW RATE

The impact of the draw solution flow rate on the PRO performance is illustrated in Figure 5. The CLPRO process testing conditions were a 26 bar hydraulic pressure and 350 m^2 membrane area. Simulation results show that the membrane flux remained constant at 19 L/m^2h with the increase of draw solution flow rate from 5500 L/h to 8000 L/h (Figure 5A). This was achieved by decreasing the draw solution inlet concentration with the increase of draw solution flow rate (Figure 5B); the draw solution inlet concentration was 1.58 mol/L at 5500 L/h but decreased to 1.49 mol/L at 8000 L/h. It is envisaged that decreasing the draw solution inlet concentration would reduce the initial chemical cost and the impact of reverse salt diffusion in the CLPRO process. Reducing the draw solution inlet concentration decreased the osmotic pressure driving force of the draw solution, as shown in Figure 5A. The osmotic pressure of the inlet draw solution was 72 bar at a 5500 L/h draw solution flow rate but decreased to 67 bar at a 8000 L/h draw solution flow rate. This reduction in the draw solution inlet concentration kept the membrane flux constant despite the increase of draw solution flow rate. However, the draw solution outlet concentration increased from 0.71 mol/L at a 5500 L/h draw solution flow rate to 0.81 mol/L at a 8000 L/h draw solution flow rate; this was mainly due to the lower membrane flux, which affected the dilution of the draw solution.

The impact of the draw solution flow rate on the CLPRO recovery rate and dilutive concentration polarization is illustrated in Figure 5C. Simulation results show that CLPRO recovery rate remained constant at 84% despite the increase of draw solution flow rate from 5500 L/h to 8000 L/h. This result was expected because %Re is the ratio of permeate flow rate to the feed flow rate and these parameters remained constant with the increasing draw solution flow rate. Simultaneously, the effect of dilutive draw solution, $e^{-J_w/k}$, did not change with the increase of draw solution flow rate due to the constant membrane flux across the PRO membrane of constant active area, i.e., 350 m^2. The practical implications of increasing the draw solution concentration at lower draw solution flow rate is realized in the operating cost; i.e., the draw solution pumping cost can be reduced by increasing the draw solution concentration. Figure 5D shows the impact of increasing the draw solution flow rate on the CLPRO power density and power generation. The power density, W, remained constant at 17 W/m^2 despite the increase of draw solution flow rate from 5500 L/h to 8000 L/h; this is approximately 3.5 times higher than the required *W* threshold for an economically viable PRO process. This was achieved because of the constant hydraulic pressure and membrane flux at different draw solution flow rates. Similarly, power generation, P_w, remained constant at the range of draw solution flow rates; 6 kW/h P_w was achieved by the CLPRO.

PRACTICAL IMPLICATIONS OF PRO OPTIMIZATION

PRO, as a renewable energy technology, can be applied for power generation in remote and landlocked regions where access to coastal areas is not possible. In such situations, CLPRO is an option because the draw and feed solutions are recycled after the regeneration process (Figure 1). CLPRO also processes advantages such as a small footprint, no loss of draw agent, controllable operating conditions, and no discharge to the environment. CLPRO can be combined with other renewable energy technologies such as solar energy for the regeneration of the draw solution by using thermal processes. The process can benefit from a waste heat source for the regeneration of draw solution; this will reduce the operating cost.

The membrane area, hydraulic feed pressure, draw solution flow rate and draw solution concentration are some of the key parameters that must be considered during the process optimization. Simulation results have shown that membrane area had only a slight effect on the CLPRO performance. However, PRO membranes with high packing density, such as Toyobo hollow fiber membranes, would reduce the capital cost of the process by reducing the number of pressure vessels of the osmotic power plant. For example, the Toyobo hollow fiber membrane has a membrane area between 500 m^2 and 700 m^2 per module [32-33]; this is a 25 to 35x larger membrane area than the HTI spiral wound FO/PRO membrane. Clearly, this can reduce not only the number of pressure vessels but also the plant footprint. Therefore, increasing the PRO membrane area is recommended firstly to reduce the capital cost and secondly to improve the PRO performance. On the other hand, hydraulic pressure has demonstrated significant impacts on the CLPRO performance. Practically, hydraulic pressure impacts the power density, which has been described as the key parameter for determining the performance of the CLPRO process. For HTI membranes, A_w equals 6.7×10^{-4} m/h. bar; a 5 W/m^2 power density can be achieved at a 14 bar hydraulic pressure, 8000 L/h draw solution flow rate and 0.73 mol/L draw solution inlet concentration. As mentioned, a 5 W/m^2 power density is the minimum threshold for an economically viable PRO process as predicted by previous studies. However, this threshold has been suggested for an open-loop PRO process which includes the pretreatment of the feed and draw solutions. Therefore, this threshold can be decreased for a CLPRO where this pretreatment is not required. The simulation results showed that increasing the hydraulic pressure from 14 bar to 26 bar resulted in a 3.5 times increase of the CLPRO power density. It should be mentioned that these simulation results refer to the maximum power density that can be achieved at $\Delta P = \Delta\pi / 2$.

Moreover, the impact of the draw solution flow rate on the CLPRO performance was evaluated at a fixed membrane area and hydraulic feed pressure. Previous studies on open-loop PRO process have shown that increasing the draw solution flow rate will increase the membrane flux and improve the process performance this will also increase the operation cost. In the CLPRO process, both the draw and feed solutions are recycled and reused, hence it is better to increase the draw solution concentration to improve the membrane flux and the process performance. This may increase the chemical cost but in the case of an NaCl draw solution, the cost of chemicals are very reasonable. When using thermolytic draw solutions the cost of chemicals could be higher; the osmotic pressure of ammonium carbon dioxide, for example, is less than NaCl, but the regeneration cost is cheaper. However, CLPRO is only feasible when there a free source of waste heat available. In such cases, the capital cost of the regeneration process is the main concern. Furthermore, the CLPRO process can be used for energy storage during off peak demand periods. Excess energy is used to separate the draw solution from the feed solution by thermal processes such as Multi Effect Distillation (MED). This energy can be recovered by mixing the draw and feed solutions using the PRO technique. This concentrated saline solution energy could be an alternative to the option of conventional energy storage technologies. The advantages of CLRPO are i) low capital cost ii) a modular system which can be scaled-up rather easily iii) low impact on the environment and marine navigation and iv) no waste discharge to the environment. To date, all PRO pilot plants have been of the open-loop design; the CLPRO design should be considered for power generation benefiting from waste heat sources or as an energy storage concept.

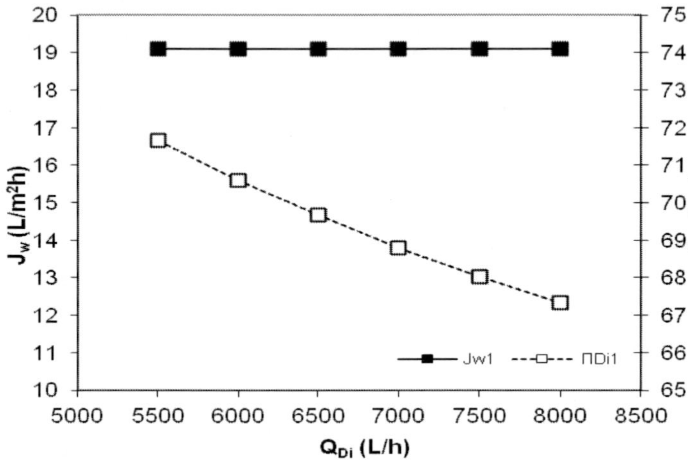

A

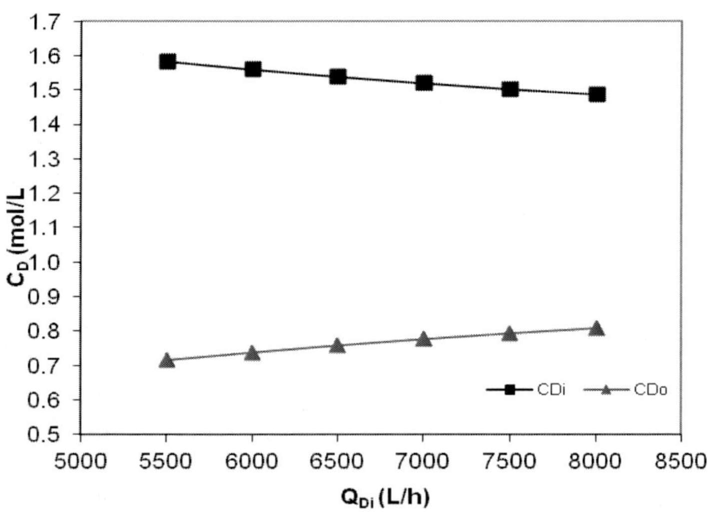

B

Figure 5 (Continued).

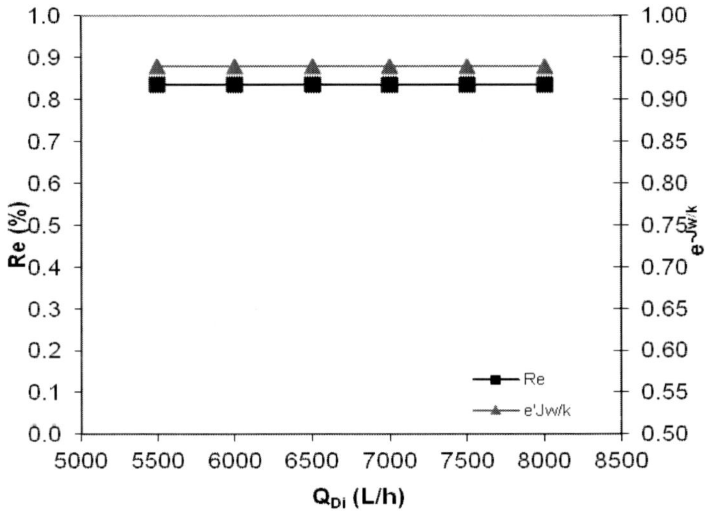

C

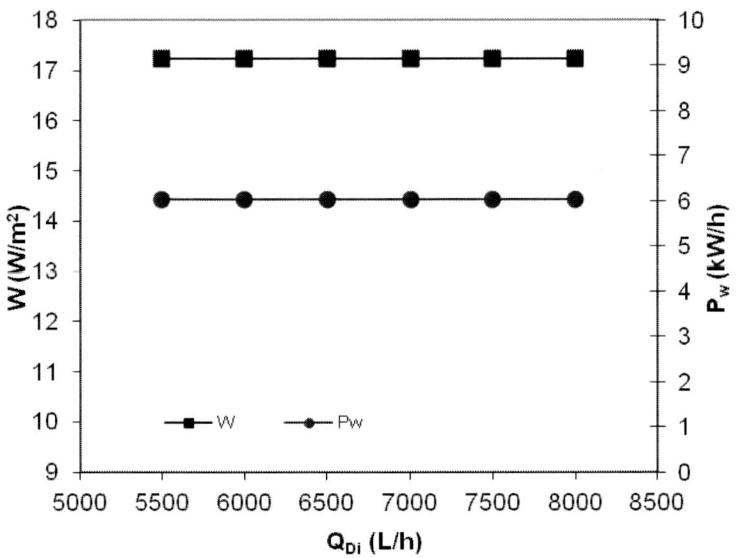

D

Figure 5. The impact of draw solution flow rate on the performance of CLPRO A) membrane flux and osmotic pressure B) concentration of draw solution C) recovery rate and dilutive concentration polarization D) power density and power generation.

CONCLUSION

The CLPRO process was suggested for power generation from a salinity gradient resource. Key operating parameters such as membrane area, hydraulic pressure, and draw solution flow rate were evaluated in the optimization of the CLPRO process. Simulation results showed that membrane area has limited impact on the performance of CLPRO; a 1% increase in power density was achieved as a result of a membrane area increase from 300 m^2 to 350 m^2. However, power density increased 3.5 times as a result of a hydraulic pressure increase from 14 to 26 bar. Considering draw solution flow rate, power density remained constant with an increase of draw solution flow rate from 5500 L/h to 8000 L/h. however, the draw solution concentration decreased from 1.58 mol/L at 5500 L/h to 1.49 mol/L at 8000 L/h to keep the membrane flux constant despite the increase of draw solution flow rate.

Power density of the CLPRO was greater than 5 W/m^2 in most cases when the hydraulic pressure was kept over 14 bar. At 26 bar hydraulic pressure and 5000 L/h draw solution flow rate, the power density remained over 17 W/m^2 despite the change of membrane area from 300 m^2 to 350 m^2. At 350 m^2 membrane area and 5000 L/h draw solution flow rate, the power density was 5 W/m^2 at 14 bar hydraulic feed pressure. This was actually the minimum threshold for an economically viable PRO process. When changing the draw solution flow rate from 5500 L/h to 8000 L/h, the power density remained approximately 17 W/m^2 at 350 m^2 membrane area and 26 bar hydraulic pressure. The results showed that the CLPRO process is economically feasible and can generate 5 W/m^2 power density provided that hydraulic pressure should be is 14 bar or greater. A pilot plant study is required to demonstrate the feasibility of CLPRO.

REFERENCES

[1] Government of Western Australia, Department of Mines and Petroleum (2013). WA Hydraulic Fracture Stimulation—Well List. *http://www.dmp.wa.gov.au/documents/well_list.pdf*, 20/05/2013.

[2] J. Nakano et al. (2012). Prospects for shale gas development in Asia: examining potentials and challenges in China and India. *Center for Strategic and International Studies*, 2012:12.

[3] PFC Energy (2011). Gazprom's pricing strategy paid off in 2010. *Memo*, 18/02/2011.

[4] Australian Government Clean Energy Regulator (2014). Carbon pricing mechanism. *http://www.cleanenergyregulator.gov.au/Carbon-Pricing-Mechanism/Pages/default.aspx*, 01/05/2014.

[5] SSY Consultancy and Research, Ltd., *SSY's Coal Trade Forecast*, 20(4) (London, United Kingdom, August 2012).

[6] A. Lewis, S. Estefen, J. Huckerby, W. Musial, T. Pontes, J. Torres-Martinez, Ocean energy, in: O. Edenhofer, R. Pichs-Madruga, Y. Sokona, K. Seyboth, P. Matschoss, S. Kadner, T. Zwickel, P. Eickemeier, G. Hansen, S. Schlömer, C. von Stechow (Eds.). IPCC Special Report on Renewable Energy Sources and Climate Change Mitigation. Cambridge University Press, Cambridge and New York, 2011.

[7] International Energy Outlook 2013. US Energy Information Administration, DOE/EIA-0484 (2013).

[8] A. E. Kontorovich, M. I. Epov, L. V. Eder (2014). Long-term and medium-term scenarios and factors in world energy perspectives for the 21st century. *Russian Geology and Geophysics*, 55(5-6):534-543.

[9] Alexander P. Mathews (2014). Renewable Energy Technologies: Panacea for World Energy Security and Climate Change. *Procedia Computer Science*, 32:731-737.

[10] R. E. Pattle (1954). Production of electric power by mixing fresh and salt water in the hydroelectric pile. *Nature*, 174:660.

[11] Stefano Mandelli, Jacopo Barbieri, Lorenzo Mattarolo, Emanuela Colombo (2014). Sustainable energy in Africa: A comprehensive data and policies review. *Renewable and Sustainable Energy Reviews*, 37:656-686.

[12] Ali Altaee, Adel Sharif (2015). Pressure retarded osmosis: advancement in the process applications for power generation and desalination. *Desalination*, 356:31–46.

[13] Willy R. Thelin, Edvard Sivertsen, Torleif Holt, Geir Brekke (2013). Natural organic matter fouling in pressure retarded osmosis. *Journal of Membrane Science*, 438:46–56.

[14] Ali Altaee, Guillermo Zaragoza, Adel Sharif (2014). Pressure retarded osmosis for power generation and seawater desalination: Performance analysis. *Desalination*, 344:108-115.

[15] Sidney Loeb (2001). One hundred and thirty benign and renewable megawatts from Great Salt Lake. *The possibilities of hydroelectric power by pressure-retarded osmosis*, 141:85–91.

[16] Ali Altaee, Graeme J. Millar, Guillermo Zaragoza, Adel Sharif. Energy Efficiency of RO and FO-RO system for High Salinity Seawater Treatment. *Clean Technologies and Environmental Policy*, DOI 10.1007/s10098-016-1190-3.

[17] Stover R. L. (2007). Seawater reverse osmosis with isobaric energy recovery devices. *Desalination*, 203:168-175.

[18] Daniel D. Anastasio, Jason T. Arena, Emily A. Cole, Jeffrey R. McCutcheon (2015). Impact of temperature on power density in closed-loop pressure retarded osmosis for grid storage. *Journal of Membrane Science*, 479:240–245.

[19] A. Altaee, A. Sharif, G. Zaragoza, N. Hilal (2014). Dual stage PRO process for power generation from different feed resources. *Desalination*, 352:118-127.

[20] Baoxia Mi, Menachem Elimelech (2010). Organic fouling of forward osmosis membranes: Fouling reversibility and cleaning without chemical reagents. *Journal of Membrane Science*, 348:337–345.

[21] Keiichiro Saito, Morihiro Irie, Shintaro Zaitsu, Hideyuki Sakai, Hidechito Hayashi and Akihiko Tanioka (2012). Power generation with salinity gradient by pressure retarded osmosis using concentrated brine from SWRO system and treated sewage as pure water. *Desalination and Water Treatment*, 41:114-121.

[22] Gang Han, Qingchun Ge, Tai-Shung Chung (2014). Conceptual demonstration of novel closed-loop pressure retarded osmosis process for sustainable osmotic energy generation. *Applied Energy*, 132:383-393.

[23] Willy R. Thelin, Edvard Sivertsen, Torleif Holt, Geir Brekke (2013). Natural organic matter fouling in pressure retarded osmosis. *Journal of Membrane Science*, 438:46–56.

[24] Stein Erik Skilhagen, Dr. Rolf Jarle Aaberg (2006). Osmotic power - Power production based on the osmotic pressure difference between fresh water and sea water. *Owemes*, 2006:20-22, Citavecchia, Italy.

[25] Natsuki Kishizawa, Koichi Tsuzuki, Masaki Hayatsu (2015). Low pressure multi-stage RO system developed in “Mega-ton Water System” for large-scaled SWRO plant. *Desalination*, 368:81-88.

[26] Akihiko Tanioka (2016). Preface to the special issue on "Pressure Retarded Osmosis in Megaton Water System Project." *Desalination*, 389:15-17.

[27] Jeffrey R. McCutcheon, Menachem Elimelech (2006). Influence of concentrative and dilutive internal concentration polarization on flux behavior in forward osmosis. *Journal of Membrane Science*, 284:237-247.

[28] Gang Han, Sui Zhang, Xue Li, Tai-Shung Chung (2015). Progress in pressure retarded osmosis (PRO) membranes for osmotic power generation. *Progress in Polymer Science*, 51:1-27.

[29] Ngai Yin Yip, Alberto Tiraferri, William A. Phillip, Jessica D. Schuiffman, Laura A. Hoover, Yu Chang Kim, Menachem Elimelech (2011). Thin-Film Composite Pressure Retarded Osmosis Membranes for Sustainable Power Generation from Salinity Gradients. *Environ Sci Technol*, 45:4360-4369.

[30] Andrea Achilli, Tzahi Y. Cath, Amy E. Childress (2009). Power generation with pressure retarded osmosis: An experimental and theoretical investigation. *Journal of Membrane Science*, 343:42–52.

[31] Karen Gerstandt, K.-V. Peinemann, Stein Erik Skilhagen, Thor Thorsen, Torleif Holt (2008). Membrane processes in energy supply for an osmotic power plant. *Desalination*, 224:64-70.

[32] Neil A. Thompson, Peter G. Nicoll. Forward Osmosis Desalination: A Commercial Reality. *Proceedings IDA World Congress*, Perth, Western Australia, September 2011.

[33] P. G. Nicoll, N. A. Thompson, M. R. Bedford. Manipulated Osmosis Applied To Evaporative Cooling Make-Up Water – Revolutionary Technology. *Proceedings IDA World Congress*, Perth, Western Australia, 4–9 September 2011.

In: Energy Efficiency
Editor: Julian E. Perez

ISBN: 978-1-53611-040-1

Chapter 5

TEMPORAL AND SPATIAL COMBINING FOR 5G MMWAVE SMALL CELLS

David Muirhead[1] and Muhammad Ali Imran[2]
[1]Honeywell Aerospace, Tewkesbury, UK
[2]School of Engineering, University of Glasgow, Glasgow, UK

ABSTRACT

This chapter proposes the combination of temporal processing through Rake combining, based on direct sequence-spread spectrum (DS-SS), and multiple antenna beamforming or antenna spatial diversity as a possible physical layer access technique for fifth generation (5G) small cell base stations (SBS) operating in the millimetre wave (mmWave) frequencies. Unlike earlier works in the literature aimed at previous generation wireless, the use of the beamforming is presented as operating in the radio frequency (RF) domain rather than the baseband domain, to minimise power expenditure as a more suitable method for 5G small cells. Some potential limitations associated with massive multiple input-multiple output (MIMO) for small cells are discussed relating to the likely limitation on available antennas and resultant beamwidth. Rather than relying, solely, on expensive and potentially power hungry massive MIMO (which in the case of a SBS for indoor use will be limited by a physically small form factor) the use of a limited number of antennas, complimented with Rake combining, or antenna diversity is given consideration for short distance indoor communications for both the SBS and user equipment (UE). The proposal's aim is twofold: to solve eroded path loss due to the effective antenna aperture reduction and to satisfy sensitivity to blockages and multipath dispersion for indoor, small coverage area base stations. Two candidate architectures are proposed. With higher data rates, more rigorous analysis of circuit power and its effect on energy efficiency (EE) is provided. A detailed investigation is provided into the likely design and signal processing requirements. Finally, the proposed architectures are compared to current fourth generation long term evolution (LTE) MIMO technologies for their anticipated power consumption and EE.

1. Introduction

In line with current thinking that 5G is unlikely to be satisfied with a 'one fits all' single air interface or access technology approach [1], specific investigation of the physical layer targeted for indoor and densely deployed SBSs warrants analysis. 5G opens the door to a range of connected technologies that are not confined to conventional mobile broadband (MBB) but may be applicable to SBSs. Application areas falling under the new category of machine type communications (MTC) will set further requirements on the air interface technology. Mission critical latencies, low energy, high reliability and high availability are new challenges in addition to high spectral efficiency and increased data rates for MBB [2]. High availability will require significant resilience to failure and with non-line of sight (NLOS) propagation being possible in mmWave, combating the effects of multipath dispersion caused by reflection is important. Examples of recent air interface research for 5G consider the use of orthogonal frequency division multiplexing (OFDM) based techniques [3, 4], the work presented here is in contrast to this and considers alternative approach based on DS-SS, specifically for the SBS.

1.1. Conventional Massive MIMO 5G Architectures

Massive MIMO or large scale antenna systems (LSAS) has been a research focus since the pivotal paper from Marzettta [5]. The premise is that the number of base station antennas N is much larger than the number of single antenna terminals, K [6]. In addition to combating eroded link margin due to reduced antenna aperture, the many antennas helps the multipath problem. When the data rate increases the symbol duration decreases. If the multipath delay spread is greater than the symbol length, complex equalization is required to combat the effects of inter-symbol interference (ISI). To overcome this, beam steering/forming is suggested to minimise the spatial footprint of the channel resulting in a more favourable multipath profile thus constraining the ISI problem. Antenna directivity is thus used to minimise the effects of ISI.

Many researchers are optimistic that the gains provided by massive MIMO will facilitate simple receiver architectures through low complexity waveforms [7] e.g., quadrature phase shift keying (QPSK). Such receivers would comprise simple matched filters, not requiring expensive and complex signal processing in the form of equalization. However, there are drawbacks. As an example, the amount of channel estimation required to formulate appropriate beamforming weights, required for either RF or digital beamforming, will be significant and will impact complexity and power. For frequency division duplex systems (FDD) this will be particularly acute. Furthermore, NLOS communication and channel blocking effects are a cause for concern. Reliability of a system purely based on beamforming might be challenging since the coverage might be more sensitive to both time and space variations [1]. Considering the spatial footprint or capture of an antenna system we refer to its beamwidth. As is discussed in [8] for an antenna configuration with a half power beamwidth (HPBW) of 6.5°, reliable communication is difficult to establish even for LOS at 60GHz and the movement of people can easily block and attenuate such a narrowbeam signal.

It is widely accepted that the use of mmWave for short indoor distances, is subject to multipath dispersion due to blockages [7, 9]. In this case NLOS communication may be the only possible means to overcome such effects. Beamforming may be inadequate in this situation if the blockage causes a significantly large attenuation and dispersion of the transmitted signal that the multipath components appear outside of the main antenna beam. Here, re-acquisition, identification and further antenna steering or beamforming may be required to find any available multipath energy.

Considering a uniform linear array (ULA) antenna array, an antenna array with uniform antenna spacing, the HPBW can be determined from:

$$HBPW = 2\left[\frac{\pi}{2} - cos^{-1}\left(\frac{1.39\lambda}{\pi N\Delta}\right)\right] \tag{1.1}$$

where Δ is the element spacing in wavelengths (λ).

A highly directive beam pattern is illustrated in Figure 1 where the polar and directivity plots of a 256 element ULA operating at 72 GHz, element spacing of 0.4 cm are provided. As can be seen a main lobe width at the HPBW point (-3 dB) from the maximum of around 1-2° is given. Considering an office or apartment indoor area of 10 m x 10 m where the maximum distance between the transmitter and receiver is 14 m $((10^2.10^2)^{1/2})$ a maximum round trip propagation delay of 93 ns (2. (dist/C)) would exist. Delay spread beyond this would be expected to be between 1 and 35 ns based on the findings in [10-12] as shown in Table 1. This would imply that multipath information may exist off the main direct path and outside the main beamwidth lobe for the 256 element ULA.

1.2. Small Cell Problem

To minimize the effects outlined, it is desirable to make use of all available multipath energy, including instantaneous multipath from blocking, and use it in a constructive way. For the case of SBS, it is desirable to use a low number of antennas to minimise power expenditure. This is indeed a constraint placed on the design not only from a power perspective but because the physical form factor limits the design to use less antenna elements. Figure 2, Figure 3 illustrate the radiation patterns more suitable for a SBS with fewer antenna elements. The plots assume the free space propagation without scatters, noise or fading. Figure 2 shows a 4 antenna ULA system operating at 28 GHz with a 1 cm element spacing. Figure 3 shows a 12 antenna ULA system operating at 72 GHz with a 0.4 cm spacing.

Table 1. Indoor mmWave Delays Spread

Freq	TX-RX Distance (m)/Area type	Delay Spread (ns)	Antenna Type
60 GHz	9.2/Modern office	20 [11]	Horn antenna
60 GHz	13.51 x 7.81/office	18.08 [12]	Omni-directional
60 GHz	13.51 x 7.81	1.05 [12]	Narrow beam
73 GHz	6-46	35 [10]	Co-polarized (15° HPBW)
73 GHz	6-46	20 [10]	Cross-polarized (15° HPBW)

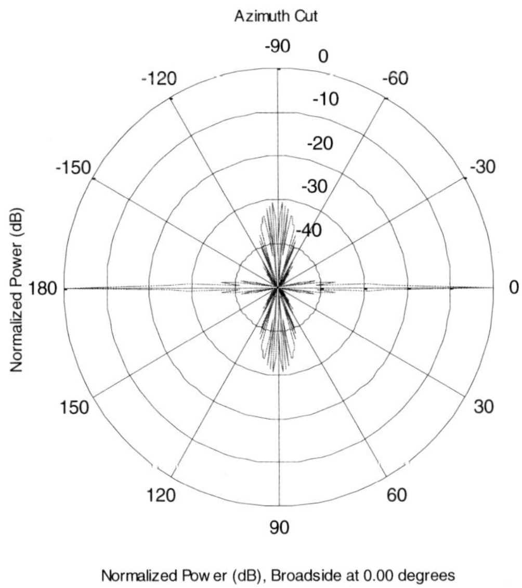

(a)

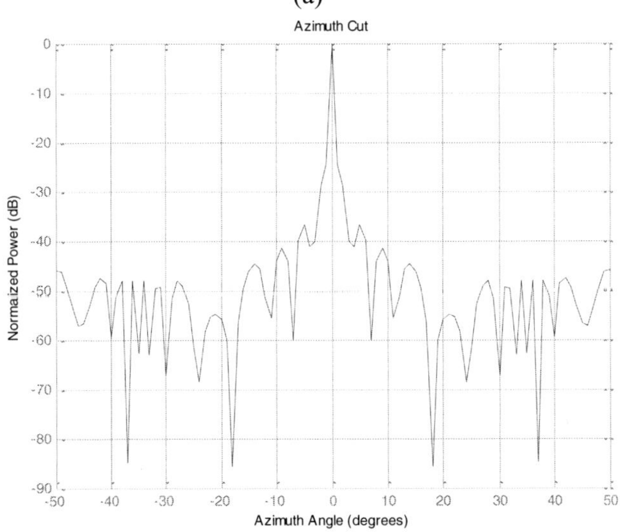

(b)

Figure 1. 256 Element, 72 GHz - (a) Polar Plot, (b) Directivity Plot.

Much wider beamwidths are evident and the spatial profile of the antenna pattern will facilitate the capture of more multipath components as well as providing moderate amounts of beamforming gain suitable for small distances. The illustrations show HPBWs of 15° and 6° for 72 and 28 GHz operation respectively. With the additional gain provided by the beamformer, the ISI problem still exists since the main lobe is sufficiently wide enough to reasonably expect multipath capture.

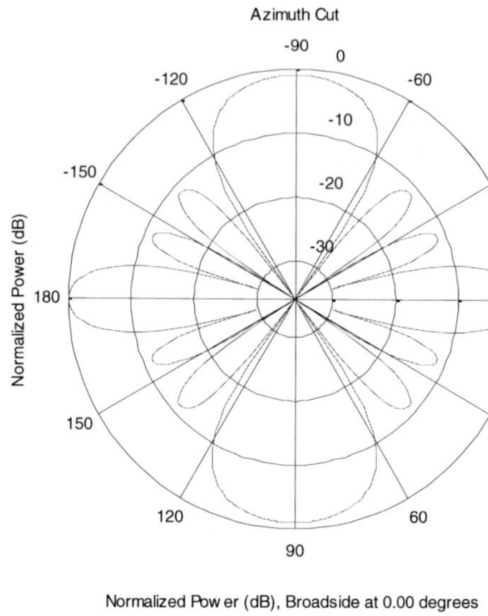

(a)

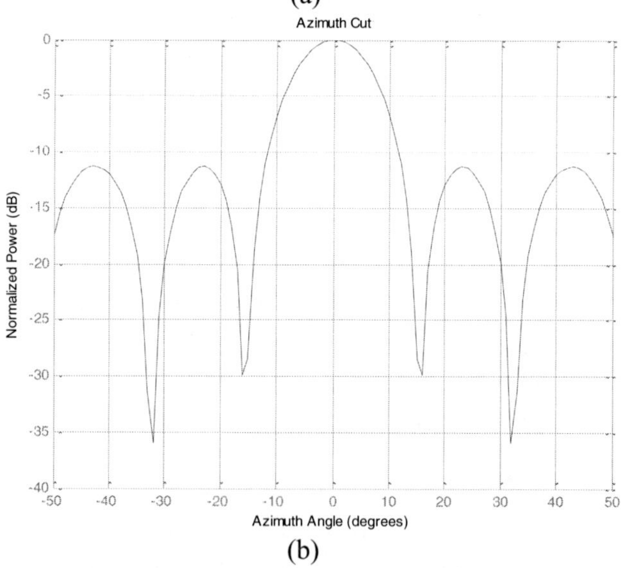

(b)

Figure 2. Polar Plot, 4 Element, 28 GHz - (a) Polar Plot, (b) Directivity Plot.

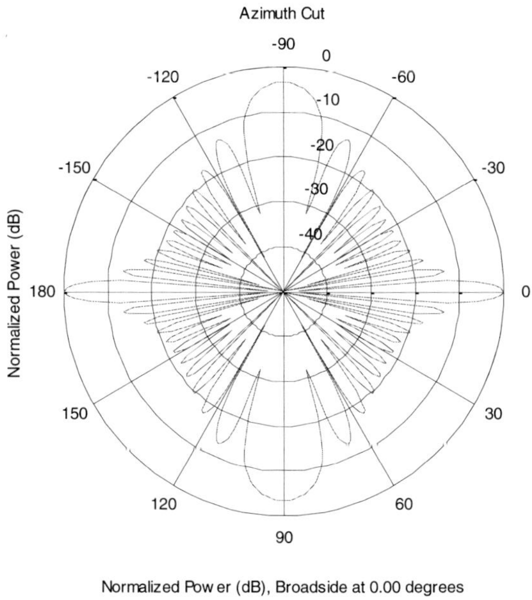

(a)

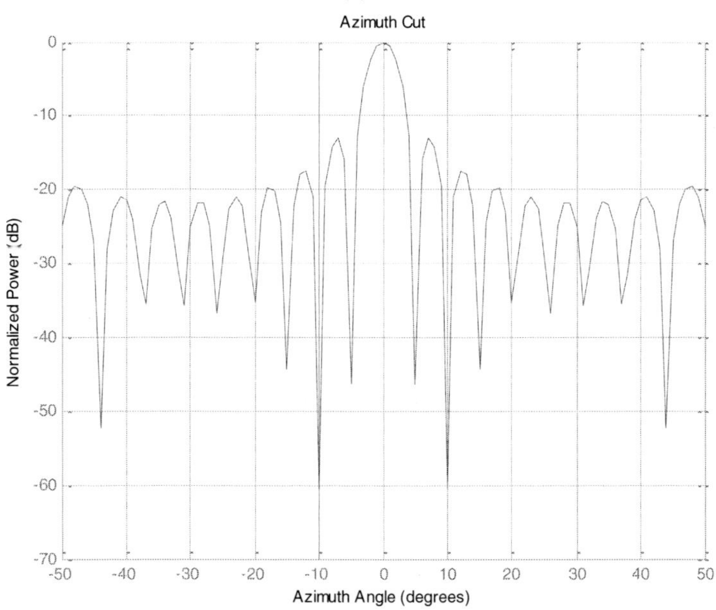

(b)

Figure 3. Polar Plot, 12 Element, 72 GHz - (a) Polar Plot, (b) Directivity Plot.

2. Small Cell Physical Layer – 2D Rake Combiner

With the requirement to overcome eroded link margin but cater for likely multipath/NLOS communications in a low complexity and energy efficient manner an alternative architecture for the small cell is considered.

The use of Rake combining related techniques such as DS-SS for mmWave has been the subject of research in the literature. This has mainly been related to indoor and predominantly for 60 GHz ultra wideband (UWB) home entertainment and multimedia systems [13]. Joint use of multiple antenna beamforming for improved link margin together with Rake combining to combat the effects of ISI, thus providing performance gains in two dimensions, are proposed and analysed. A simplified architecture is proposed to satisfy the needs of the small cell. Research as recent as June 2015 advocates the use of Rake combining techniques for indoor 60 GHz operation [14]. However, such systems, despite claiming simple and low complex solutions, do not fully quantify the extent of this in terms of circuit power and EE. As part of the discussion presented here, a more quantifiable complexity and EE performance is given.

2.1. Direct Sequence-Spread Spectrum Rake Combining and mmWave

It is well known that in direct sequence-spread spectrum (DS-SS), information is transmitted using a wider bandwidth than necessary. Here data bits from the source $b(t)$ are multiplied by a code signal $c(t)$ at a faster rate known as the chip rate. The process is referred to as spreading and the amount of spreading is determined by the spreading factor (SF). At the receiver of such a system, the use of a Rake Receiver effectively performs the equalization task by combining the received signal with multiple, time delayed versions of it (the multipaths) separated by multiples of the chip period (T_c). This allows the components of the original signal to be recovered with the time delays removed. Maximum ratio combining (MRC) of the delayed signals yields an optimal output which has a maximum possible signal to noise ratio given the input signals [15]. Optimal performance is expected because the increased bandwidth of the spread signal allows the receiver to resolve multipath energy which would otherwise appear combined with a single channel tap in other systems. In this sense the spreading mechanism gives us the ability to find available energy and use it to our advantage. In particular, considering the spread signals with chip time T_c inversely proportional to the spreading bandwidth; in this case, the individual paths can be distinguished if they are mutually separated by delays greater than T_c. The various delayed versions of the signal will be mutually nearly uncorrelated [14]. Since the mmWave frequency bands permit the use of high bandwidth spreading, the corresponding multipath resolution will be significant (sub-nanosecond). This therefore increases the effective diversity order and much, if not all, available energy can be identified and used to increase the signal to noise ratio (SNR) of the signal.

To summarize, in most receiver design multipath creates inter symbol interference that requires potentially complex equalization techniques. In the case of DS-SS such multipath components can be used in a constructive way to improve the performance of the system by detecting and combining the main and delayed multipath components in a Rake Receiver. In this case multipath actually provides additional improvements in the system performance and resilience to NLOS communications. This therefore satisfies the requirement of using all available energy and addressing ISI in the likely transmission conditions of a 5G SBSs.

2.2. Spatial and Temporal Processing

Antenna arrays at the SBS can be utilized to form spatial domain beam patterns for both the UL and DL. This maximises the desired user's signal. DS-SS provides temporal diversity gain through multipath combining. Thus, there are two dimensions, spatial and temporal, to increase the performance of the communications link in the proposed scheme. In operation the received signal is first cross-correlated with a local copy of the spreading sequence (combined with a scrambling code) such that the chip positions corresponding to the time delays of the required user can be identified [16]. A Rake combiner is used to equalise the main and time delayed multipath components to maximise the signal to interference ratio (SIR) of the signal. The use of beamforming performed at baseband will result in excessive power consumption because of the necessity to use analogue to digital converters (ADC) behind each antenna element. In the following discussion the use of beamforming is assumed to be performed in RF domain followed by correlation and other functions in the baseband domain.

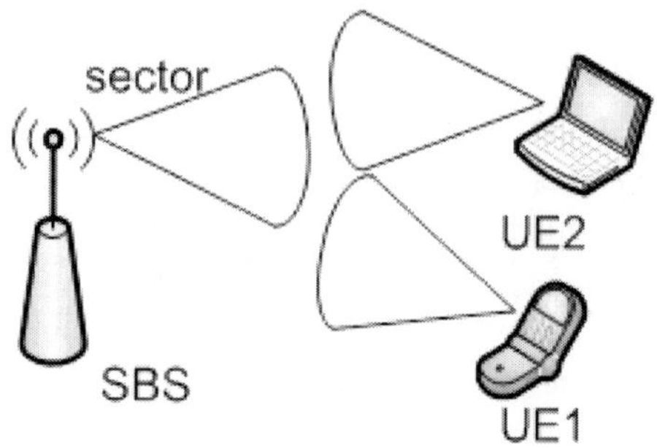

Figure 4. DL sectored and UL beamforming at SBS and UE.

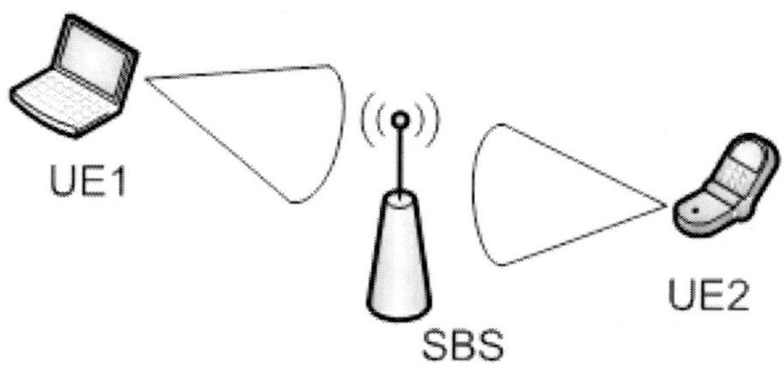

Figure 5. DL and UL antenna diversity at SBS for 360° coverage.

3. Transceiver Architectures

Since each user, in a DS-SS system is transmitted and received simultaneously at the SBS, (and identified by their unique scrambling code), the use of RF beamforming is potentially limited to providing the same antenna pattern for each user. In this case the beamforming gain can be used to sectorize a particular coverage area – see Figure 4. For the UE, RF beamforming offers a mechanism to improve the link margin by increasing the effective antenna aperture of the receiver, as well as the facilitation of transmit beamforming gain.

An alternative architecture, providing 360° coverage at the SBS, is to employ transmit and receive diversity techniques, rather than beamforming, as illustrated in Figure 5. Gain in the form of diversity gain is provided in the DL and UL. Both BS and UE receive will employ digital antenna diversity with Rake combining. Two transceiver architectures are considered and shown in

Figure 6 and Figure 7 - Architecture 1 and 2. Both use DS-SS, but the utilisation of the antenna system is different. Architecture 1 considers the use of an RF beamforming antenna array at the SBS and an antenna array providing beamforming in the RF domain at the UE. Architecture 2 improves the transmit performance and receive gain of the SBS by employing transmit and receive diversity at the expense of additional ADC/DACs. Both architectures comprise identical spreading, scrambling, filtering, and necessary up conversion and amplification in the transmit path. For Architecture 1, receive beamforming is provided across an N element array (N = 2:4). Following the beamformer, amplification, down-conversion and corresponding descrambling and despreading is provided. Multipath search estimation is used to determine appropriate multipaths in the channel that are then combined in a Rake combiner. The identification and acquisition of multipath components is required to exploit their energy and it is particularly important to determine the relative delays and when possible their amplitude and phase components.

Identification of multipath components in a transmission can be performed using the time domain correlator, implemented as a code matched filter (CMF), who's taps are equivalent to a local and known spreading and scrambling code sequence multiplied by known pilot symbols (for channel estimation). Assuming a single beamformed lobe, multipath search estimation complexity is determined by the number of expected multipaths or more precisely the timing uncertainty over which to search. In a small cell this should be low since it is constrained by the physical coverage area. A Rake combiner is used to combine the multipath components identified by the correlator. To summarise, the motivation and advantages of such a system are as follows:

1. Spreading allows the identification of time delayed multipath components by correlating the transmitted signal with a local copy of the spreading code (which will likely be combined a scrambling sequence and known pilots). In particular and with high bandwidth associated with mmWave spectrum, multipath resolution to sub-nanosecond accuracy is possible.
2. Unlike other systems, the multipaths in DS-SS can be used to perform constructive combining at the receiver with a Rake combiner. This therefore provides additional gain in the form of diversity gain and provides resistance to blocking and multipath dispersion.
3. Assuming that the number of users is limited in the small cell, the DS-SS system, which is normally limited by the maximum number of

users will be less affected by multiple access interference (MAI) and noise rise.

4. Large bandwidth exists in the mmWave bands to facilitate the additional spectrum needed by the spreading process.
5. The spreading and scrambling processes can be considered as low complexity. Arguably the most complex part is the time domain correlator but since its complexity is driven by delay spread this will be minimal in an indoor deployment where the coverage area and therefore the timing uncertainty will be small.
6. DS-SS with low order modulations such as QPSK will facilitate low peak to average power ratio (PAPR) compared to more complex modulation envelopes or multi-carrier systems (such as OFDM). This will result in more efficient use of power amplifiers.
7. A small cell radius could enable the use of lower power ADC/DAC through lower dynamic range requirements.

The disadvantage with such a scheme is the effective waste of bandwidth due to the spreading process i.e., for every symbol transmitted SF x system bandwidth is needed. Additionally, due to the higher effective signal bandwidth the ADC and DAC components will be required to clock yet higher meaning greater power consumption. In general it is preferred that the ADC has large spurious free dynamic range (SFDR) meaning large order ADC devices e.g., 12-bit [17]. However, to reduce power consumption the use of smaller dynamic range ADCs could be suitable for short distances. Unlike OFDM systems, DS-SS is a single carrier system meaning it has a lower peak-to-average power ratio (PAPR). Using low order modulation schemes such as binary phase shift keying (BPSK) or QPSK would facilitate the use of ADCs with a smaller dynamic range.

3.1. Performance Aspects

The link budget of a small cell based on the proposed 2D-Rake physical layer architecture 1 is shown in Table 2 and Table 3. Spreading factors of SF = 4 and SF = 8 are shown respectively. Particularly for the SF = 4, the achievable data rates are comparable with those discussed in [18] which considers a single omni-directional antenna but due to the spreading and multipath combining, should provide better mitigation to blocking and the effects of channel.

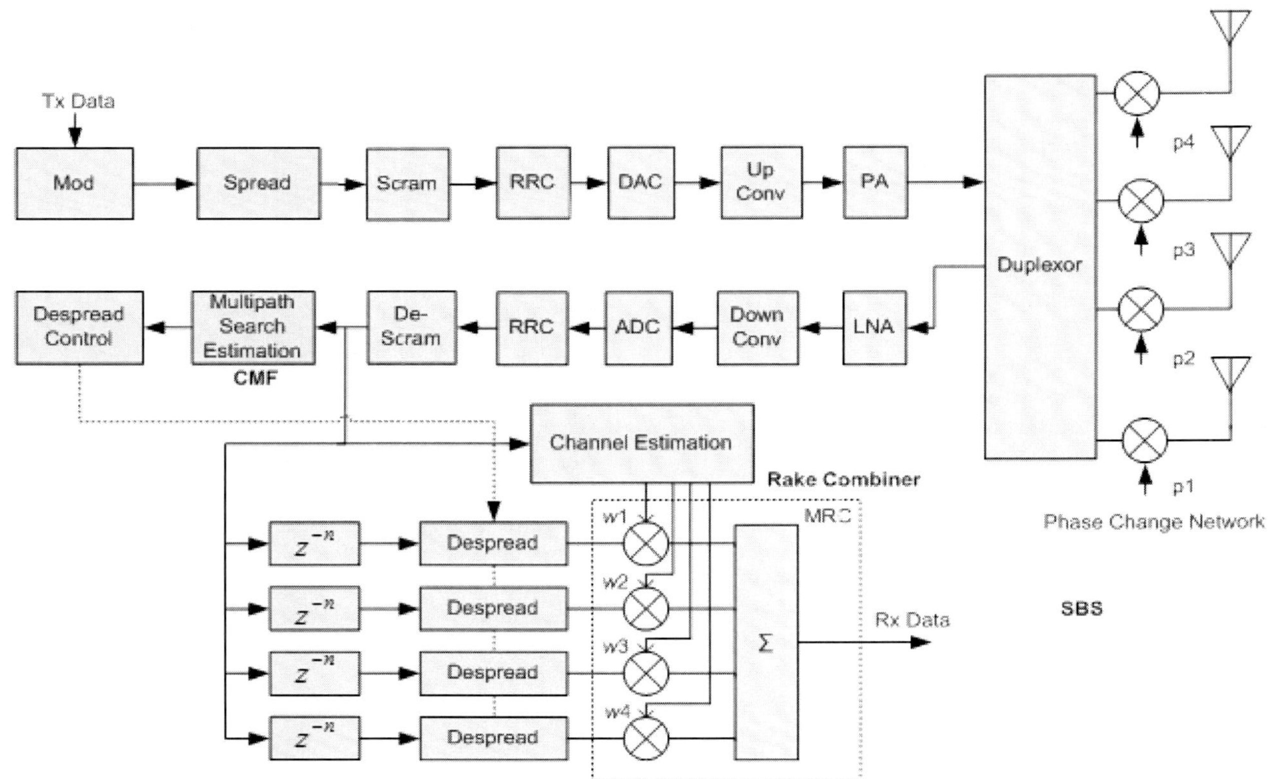

Figure 6. (Continued).

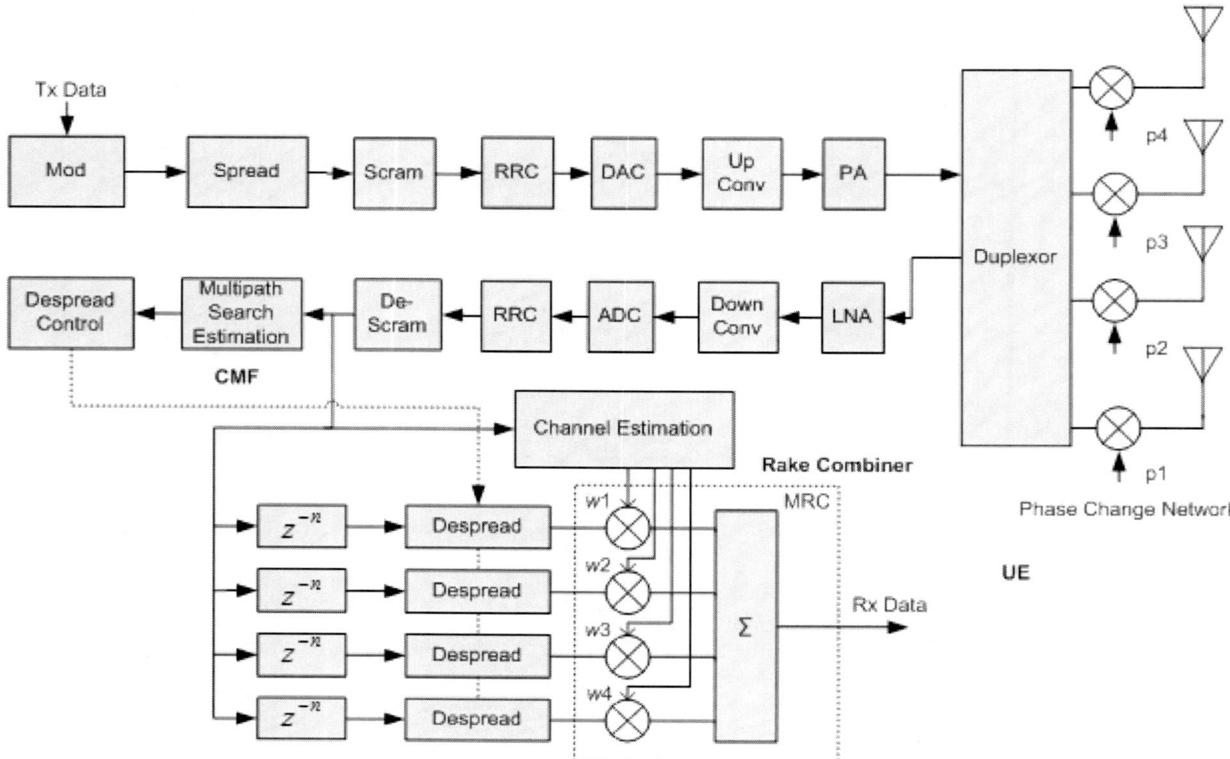

Figure 6. Architecture 1.

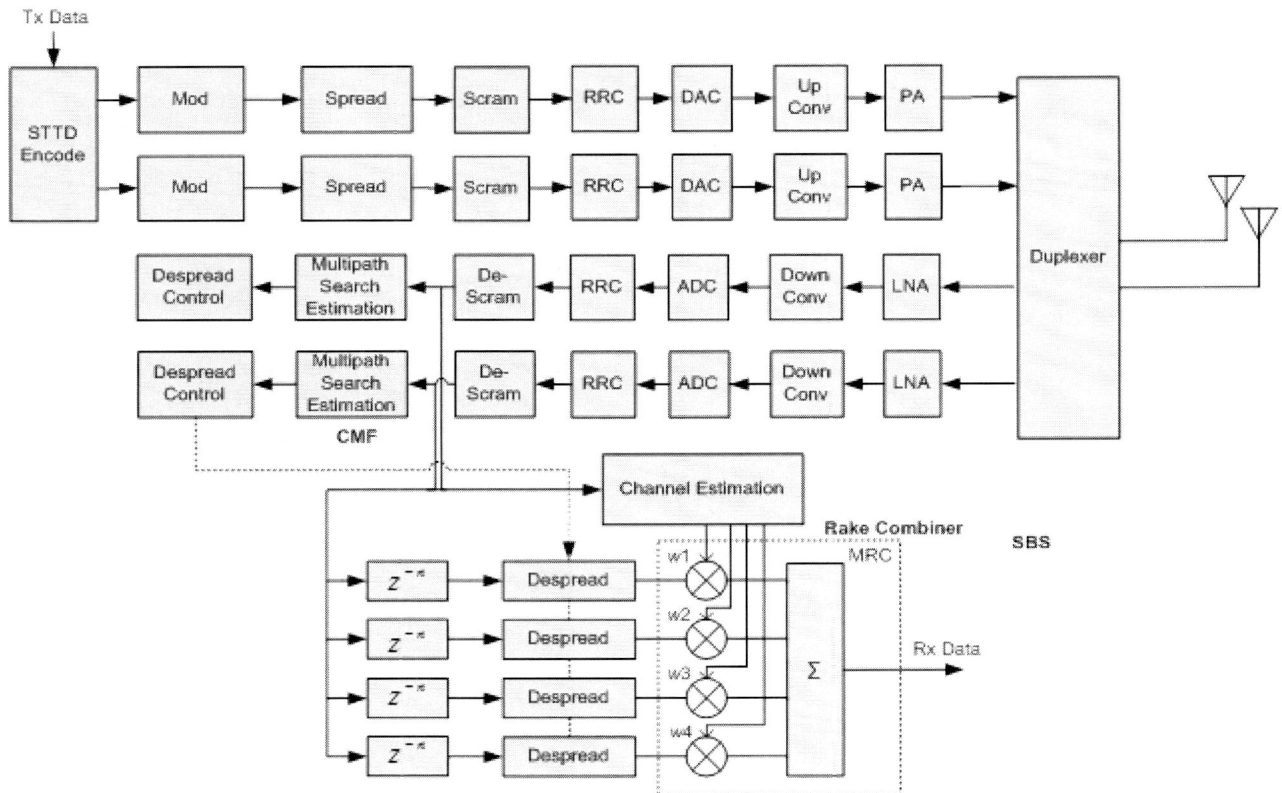

Figure 7. (Continued).

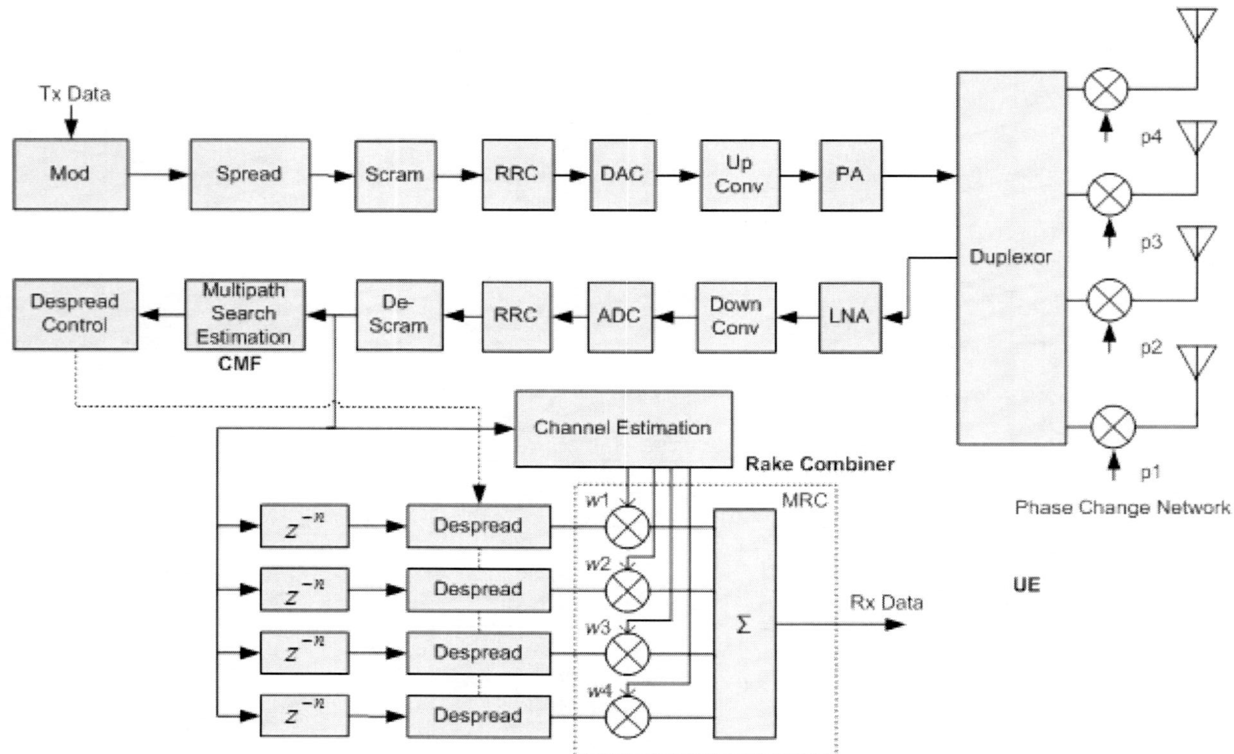

Figure 7. Architecture 2.

Table 2 and Table 3 show the achievable data rates when considering short distances (5 and 10m) between the transmitter and receiver. The data rate achievable are significant with only a four antennas at the receiver and a single transmit antenna element.

Table 2. 2D-Rake Link Budget - SF4, 5 and 10 m distance, 28 and 72 GHz

Small Cell Link Budget	Case 1 sf4	Case 2 sf4	Case 3 sf4	Case 3 sf42
Tx Power (dBm)	20	20	20	20
Beamforming Gain (dBi)	12	12	12	12
Carrier Frequency (GHz)	2.80E + 10	2.80E + 10	7.20E + 10	7.20E + 10
Distance (m)	10	5	10	5
Propagation Loss (dB)	81.39688885	75.37628893	89.60037815	83.57977824
Spreading Factor	4	4	4	4
Other Losses	6	6	6	6
Received Power (dBm)	-55.39688885	-49.37628893	-63.60037815	-57.57977824
Bandwidth (GHz)	2.00E + 09	2.00E + 09	2.00E + 09	2.00E + 09
Thermal PSD (dBm/Hz)	-174	-174	-174	-174
Noise figure	10	10	10	10
Thermal Noise (dBm)	-7.10E + 01	-7.10E + 01	-7.10E + 01	-7.10E + 01
SNR (dB)	2.16E + 01	2.76E + 01	1.34E + 01	1.94E + 01
Implementation Loss (dB)	3	3	3	3
Data rate (bits/s)	3.10E + 09	4.09E + 09	1.79E + 09	2.75E + 09

Table 3. 2D-Rake Link Budget – SF8, 5 and 10 m distance, 28 and 72 GHz

Small Cell Link Budget	Case 1 sf8	Case 2 sf8	Case 3 sf8	Case 4 sf8
Tx Power (dBm)	20	20	20	20
Beamforming Gain (dBi)	12	12	12	12
Carrier Frequency (GHz)	2.80E + 10	2.80E + 10	7.20E + 10	7.20E + 10
Distance (m)	10	5	10	5
Propagation Loss (dB)	81.39688885	75.37628893	89.60037815	83.57977824
Spreading Factor	8	8	8	8
Other Losses	6	6	6	6
Received Power (dBm)	-55.39688885	-49.37628893	-63.60037815	-57.57977824
Bandwidth (GHz)	2.00E + 09	2.00E + 09	2.00E + 09	2.00E + 09
Thermal PSD (dBm/Hz)	-174	-174	-174	-174
Noise figure	10	10	10	10
Thermal Noise (dBm)	-7.10E + 01	-7.10E + 01	-7.10E + 01	-7.10E + 01
SNR (dB)	2.46E + 01	3.06E + 01	1.64E + 01	2.24E + 01
Implementation Loss (dB)	3	3	3	3
Data rate (bits/s)	1.80E + 09	2.30E + 09	1.13E + 09	1.62E + 09

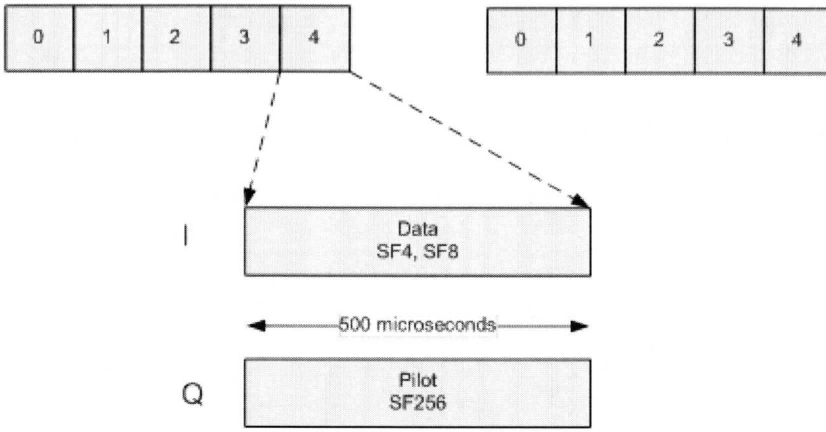

Figure 8. Proposed Slot Format - DL and UL.

3.2. Design and Complexity Aspects

A simple physical layer is suggested using QPSK modulated DS-SS. Using the 5G use case of sub-millisecond latency [19], the transmissions are split into 500 μs timeslots as shown in Figure 8. Each timeslot comprises data and pilot bits mapped to the real (I) and imaginary (Q) components of the signal respectively. Pilot symbols are used for parameter estimation of multipath components and channel state information (CSI) for appropriate weighting and receive functions.

A maximum of 4 users can be served by the SBS when the SF = 4, or 8 users when the SF = 8. Each user, u, is assigned a pseudo noise (PN) sequence of N_c chips defined as:

$$c_u[n], n = [0; 1: \ldots . N_c - 1] \tag{1.2}$$

$c_u[n]$ is the nth chip where elements of c_u are +/- 1. Assuming the system is sampled at the chip rate, transmission is represented as:

$$x_u[n] = s_u[i]c_u[n\{modN_c\}] \tag{1.3}$$

where $s_u[i]$ is the ith information bearing signal, lasting for N_c chips. The signal in the presence of noise is received as:

$$r[n] = \sum_u \sqrt{P_u} \sum_l h_u[n - \tau_{u,l}]\, x_u[n - \tau_u - l] + \omega[n] \quad (1.4)$$

where $h_u[n, l]$ = user u's channel response. τ_u is the integer delay of user u and $\omega[n]$ is additive noise.

3.2.1. Coherence Time

The coherence time ($T_{coh} = \frac{C}{f_c v} = \frac{1}{f_m}$) tells us tells us the time that the channel remains constant due to the effects of Doppler (f_m). Assuming carrier frequencies (f_c) of 28 GHz and 72 GHz with a maximum UE velocity (v) in the small cell of 0.5 km/h, coherence times of 77 ms and 30 ms respectively are given. This tells us that the channel will remain constant over the duration of many 500 μs slot periods which will help to minimise the amount of CSI required.

3.2.2. Channel Parameter Estimation – Proposed Algorithm

Both proposed architectures will require estimation of CSI (channel estimation) and significantly estimation of the multipath delay components (multipath search estimation). In both transmit and receive, data and pilot are separately BPSK modulated on the I and Q channel respectively, thus forming a QPSK constellation. Orthogonal variable spreading factor (OVSF) codes can be used on the data and pilot symbols to maintain orthogonality. Different users use different UL scrambling codes to create a unique traffic channel. Neglecting mutual interference between the data and pilots, the signal received at the base station can be separated into I and Q channels. The Q channel can be used to estimate the channel parameters.

UL signals from the users are received by an *N*-element antenna array. With perfect instantaneous power control, to ensure all users exhibit equal power, an equivalent complex baseband expression of the composite received vector **X**(*t*) at time *t* is given by:

$$x(t) = \sum_{K=1}^{K} \sum_{l=1}^{L_k} \alpha_{k,l}^{u}\, e^{j\phi_{k,l}} \sum_{n=-\infty}^{\infty} b_k(n) c_k(t - nT_b - \tau_{k,l})\, \overrightarrow{a_k}(\theta_{k,l}) + \mathbf{n}(t) = \sum_{K=1}^{K} \sum_{l=1}^{L_k} \widehat{a_k}(\theta_{k,l}) \sum_{n=-\infty}^{\infty} b_k(n) c_k(t - nT_b - \tau_{k,l}) + \mathbf{n}(t) \quad (1.5)$$

where it is assumed that there are *K* users (one desired and *K*-1 interfering users) the kth user has L_k propagation paths. The parameters $\alpha_{k,l}^{u}$, $\theta_{k,l}$, $\tau_{k,l}$ and $\theta_{k,l}$ are the UL amplitude, phase shift, time delay and angle of arrival of the *l*th multipath component respectively, from the kth user. $b_k(n)$ is the nth bit value,

$c_k(t)$ is the spreading waveform assigned to the kth user, and T_b is the bit period. Assuming Architecture 1, the column vector $a_k(\theta_{k,l}) = [1, a_1(\theta_{k,l}),\ldots, a_{M-1}(\theta_{k,l})]^T$ is the array response vector corresponding to the path arriving on angle $\theta_{k,l}$, where $a_m(\theta_{k,l})$ is a complex number denoting the amplitude gain and phase shift of the signal at the $(n+1)^{th}$ antenna relative to that at the first antenna $\overrightarrow{a_k}(\theta_{k,l}) = \alpha_{k,l}^{u} e^{j\phi_{k,l}} \overrightarrow{a_k}(\theta_{k,l})$ is the channel vector and $\mathbf{n}(t)$ is the additive white Gaussian noise vector.

$$P_{r} = \sum_{l=1}^{L_k} \left(\alpha_{k,l}^{u}\right)^2, for\ all\ K \quad (1.6)$$

P_r, is the total received power from the kth user and is assumed to be constant for all because of perfect instantaneous power control.

Following the beamformer, multipath search and estimation is performed. Assuming that the first user is the desired user and code synchronization has been established, the output of the, CMF results in the power delay profile of the desired user k_1) is given as [16]:

$$\mathbf{z}_1(\tau) = T_b b_1(n) \sum_{l=1}^{L_1} \widehat{a_k}(\theta_{1,l})\delta(\tau - \tau_{1,l}) + s_1(\tau) + \mathbf{m}_1(\tau) + \eta_1(\tau) \quad (1.7)$$

where s_1 is the self-interference signal vector due to other multipath components of the desired user, $\mathbf{m}_1$ is the MAI vector, and η_1 is the thermal noise vector. The output of the matched filter is used to distinguish the desired signal from the co-channel interference who's time resolution is the chip interval or fraction of the chip interval T_c, depending on the oversampling factor.

The complexity of the CMF, assuming a parallel search of all timing uncertainties are searched for within the period of a chip, which is the optimal solution, is determined by the maximum delay spread searched over in nanoseconds. Assuming the delay spreads discussed in section Table 1 an average delay spread of 18 ns can be expected for an indoor small cell.

Figure 8 shows that data is transferred in multiple timeslots where each slot comprises N pilot bits to aid channel and delay estimation and consider a slow fading channel. The delay profile for each pilot is first obtained, under the assumption that a long scrambling code is used. It is also noted that during a time slot period, the total phase change of the desired signal due to Doppler shift is small, if we assume a UE speed of 1 km/h, and the slot period of $T_{slot} = 500\mu s$ ($\theta = f_d.\ 360°.\ T_{slot}$). This enables the use of coherent accumulation of the pilot chips, to obtain the mean delay profile of the user. With this approach, the mean delay profile, $\bar{z}_1$, is given by:

$$\bar{z}_1(\tau) = \frac{1}{N}\sum_{n=1}^{N} z_{1,n}(\tau) \tag{1.8}$$

where $z_{1,n}(\tau)$ is the CMF output of the nth pilot bit. Coherent integration will be applied to the time bins corresponding to the delay times of the desired user's multipath and will be accumulated N times [16]. In addition to coherent integration, and to further reduce the MAI a, non-coherent integration of the absolute value of the mean delay profile, $\bar{z}_1$, over the S timeslots is performed as follows:

$$\bar{\bar{z}}_1(\tau) = \frac{1}{S}\sum_{s=1}^{S} |\bar{z}_1(\tau)| \tag{1.9}$$

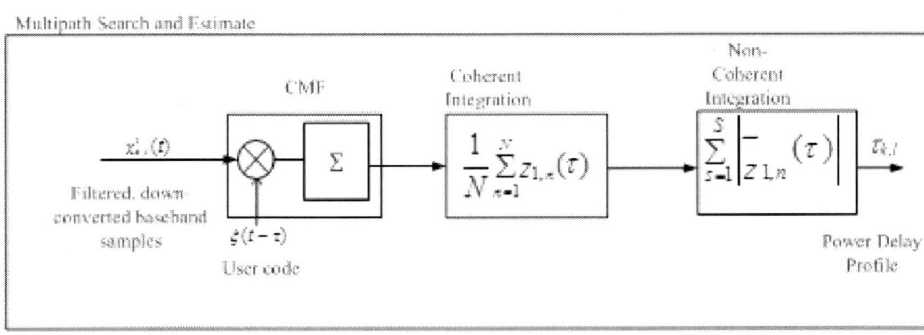

Figure 9. Multipath Estimation.

3.2.3. Multipath Search Estimation - Complexity

Determination of complexity can be broken down into two distinct timing acquisition functions:

- UE timing acquisition
- Multipath timing detection

Both these functions are determined by cross correlating the incoming received chips, or fractions of chips, with a local copy of the spreading and scrambling code sequence. The operations involve sign-bit complex correlations. The number of these operations, so called hypotheses [15], is determined by the round trip delay between the transmitter and receiver and the delay spread. These are therefore minimised in a small cell where the round trip delay is proportional to the cell size in the order of metres - $d_{rt} = T_c C/2$, where d_{rt} is the round trip delay distance in metres corresponding to T_c, the chip duration. Multipath delay spread can be assumed to be an average of 18 ns based on Table 1. Considering a chip rate of 2Gcps, the round trip delay incurred by each chip would be 0.15 m. With a maximum coverage of 10 m, a timing search with timing uncertainty of 133.33 chips would be required. Beyond this an additional 18 ns of delay spread needs to be accounted for which results in a complex correlator of 169 taps ($T_{\mathrm{d}} = 84\, ns$).

The timing acquisition process involves testing all likely hypotheses. Traditionally, in large cell coverage areas, resource limitations meant that the hypotheses testing would be performed in a serial manner where each incorrect hypothesis is eliminated before the next one is tested [15]. Performing the operation in a parallel fashion increases the complexity meaning that for every chip, or fraction of a chip, received all possible timing uncertainties could be tested at once i.e., for the timing uncertainty above, 169 complex correlation operations would be made in 1 chip duration, $T_c = 0.5$ ns. The UE speed will also have an impact on how the timing acquisition is performed. Considering a UE in a small cell with a maximum velocity of 1 km/h, a maximum movement per chip period, T_c, would be in the order of 13.5 nm (13.5 nm/chip) meaning an equivalent chip fraction of ($T_c C$/13.5e-9) ~ = 11.1e6 times smaller. In other words it would take 0.0055s for the timing uncertainty to move by 1 chip. A minimum search update frequency of 180 Hz would therefore be required.

4. Energy Efficiency of the Proposed Architectures

At face value, the 2D rake system appears low in complexity in terms of baseband operations. For example the spreading and scrambling operations include single bit modulo-2 add operations, as shown in Figure 10.

With the expected Gbits/s data rates anticipated for 5G systems, the circuit power P_C will play an increasingly important role in determining the energy efficiency of applicable architectures. As the bandwidth increases, so does the internal clocking of digital signal processor and associated digital and analogue hardware, leading to an increase in power usage. It is therefore prudent to rigorously consider circuit power consumed by signal processing.

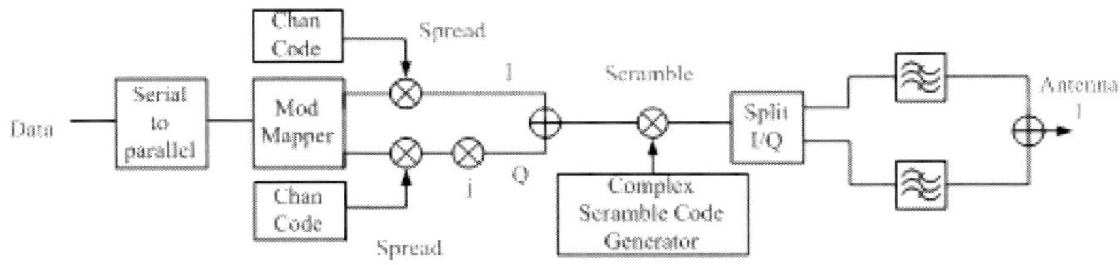

Figure 10. Spreading and Scrambling Operations.

As discussed in [21] theinternal power, P_c, comprising baseband computations and A/D and D/A conversions and is given as :

$$P_c = \frac{2R_{flops}}{a} + Mb \tag{1.10}$$

where $R_{flops,}$ is the total computational rate in floating point operations per second (FLOPS), a is the power efficiency of computing measured in flops/watt given as 12.8 GFLOPS/W [21]. M is the number of antennas and b is the internal power consumption associated with each antenna. Since there is uncertainty as to how much internal power is used a wide range of values of b can used [21]. b (in mW) = 32, 64, 128, 256, 512, 1024, 2048, 4096.

For both architecture 1 and 2, it can be assumed that the critical computations will be the spreading and scrambling in the DL. Whereas in the UL, descrambling, despreading and complex correlation operations (for multipath timing acquisition) will dominate. As in OFDM, the 2D-Rake will be dominated by the number of complex multiplications. However, in practice many of these will include single bit multiplicands, for example modulo-2 additions for spreading. Assuming all operations are treated as full word size complex multiplies, an equivalent DL complexity for the 2D-Rake slot, per SBS j, can be given as:

$$R_{\text{flops},j,DL} = M_j \left[K_J \frac{T_{\text{slot}}}{T_{\text{c}}}\right] \tag{1.11}$$

and for the UL:

$$R_{\text{flops},j,UL} = M_j \left[K_J \left(\frac{T_{\text{slot}}}{T_{\text{c}}} + \frac{T_{\text{p}}}{T_{\text{c}}}\frac{T_{\text{d}}}{T_{\text{c}}}\right)\right] \tag{1.12}$$

where the UL operations are a combination of despreading together with timing uncertainty search functions associated with the multipath estimation process shown in Figure 9. Equation (1.12) assumes the use of a optimul implementation using a parrallel CMF.

The rate at which the multipath detection is performed does not need to be continual. To provide the appropriate power consumed in 1 second, equation (1.12) can be extended to include a specified search rate R_s in Hz.

$$R_{\text{flops},j,UL} = M_j \left[K_J \left(\left(\frac{T_{\text{slot}}}{T_c} \right) 1/T_{\text{slot}} + \left(\frac{T_p}{T_c} \frac{T_d}{T_c} \right) R_s \right) \right] \quad (1.13)$$

Referring to the relationship between SE and EE expression (1.14)

$$\eta_{EE} = \frac{W\eta_{SE}}{P_c + N_0 W (2^{\eta_{SE}} - 1)/\rho} \quad (1.14)$$

the impact of the circuit complexity can be determined. Figure 11 shows the SE-EE trade-off as a function of the search rate for both UL and DL processing of the SBS (Architecture 1). As previously indicated, and due to the relative slow movement of the indoor users, the multipath estimation process does not require continual updates. The graph shows dramatic increases in EE when the rate is reduced to 180 Hz or 50 Hz where a significant resultant lower power is achieved - (Architecture 1, BW = 1 GHz, b = 32 mW, a = 12.8 GFLOPS/W, PA efficiency =25%, number of users =4, number of baseband paths =1).

An equivalent graph for Architecture 2 is shown in Figure 12 and assumes the use of transmit diversity in the DL and receive diversity in the UL. In this case the baseband complexity is doubled as is the internal, non-RF generating circuitry power, b. (Architecture 2, BW = 1 GHz, b = 32 mW, a = 12.8 GFLOPS/W, PA efficiency =25%, number of users =4, number of baseband paths =2).

4.1. Comparison with 4G

To determine the suitability of the 2D-Rake architectures for their power and energy efficiency, it is useful to make comparisons with current fourth generation base stations, based on the LTE standard which comprises OFDM techniques on the DL and UL at the BS. With reference to the material presented in [20], appropriate transmit and received baseband operations and power can be determined for an LTE based femtocell base station (FBS), a

SBS sub-class.. The authors assume a 40GOPS/W as the reference figure to determine the power (P) used. This is based on both the base station type and the underlying silicon feature size/geometry which was 65 nm in 2010. The figure is scaled up 3 times for the FBS based on the assumption that more power efficient dedicated hardware would be used. An illustrative LTE block diagram is given in Figure 13 which shows the main baseband components. In addition, a halving of total dynamic power is attributed to a change from 65 nm to 45 nm complementary metal oxide semiconductor (CMOS) technology.

Table 4. 2D-Rake Operating Parameters

K_J: Users simultaneously served per SBS	4, 8
M_J: No. of antennas per SBS	2, 4
T_{slot}: Slot length	500 μs
T_c: Chip Interval	0.5 ns
T_P: Pilot Interval	500 μs
T_d = Timing uncertainty (dependent on the cell size and delay spread)	84 ns

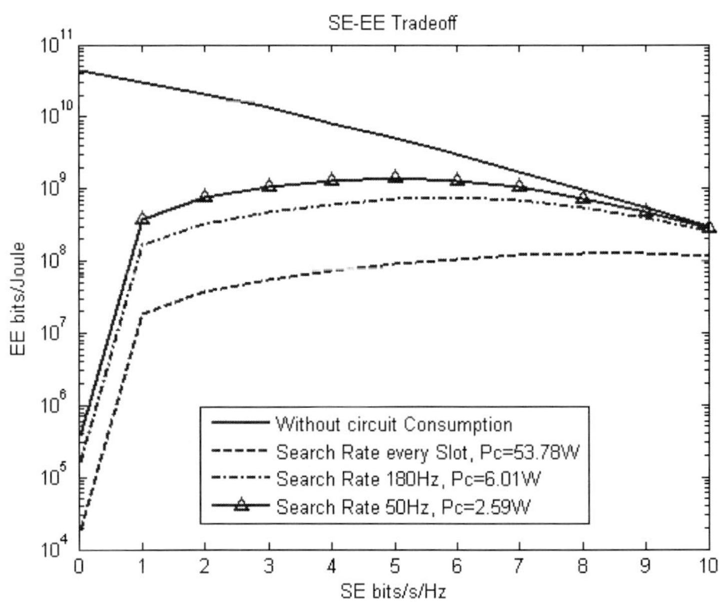

Figure 11. 2D-Rake SE/EE trade-off as a function of Search Rate - Architecture 1.

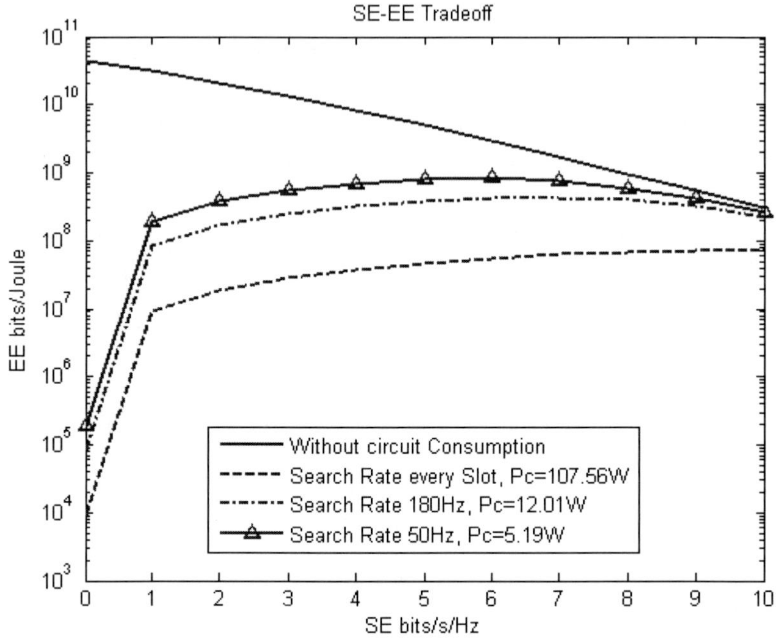

Figure 12. 2D-Rake SE/EE trade-off as a function of Search Rate - Architecture 2.

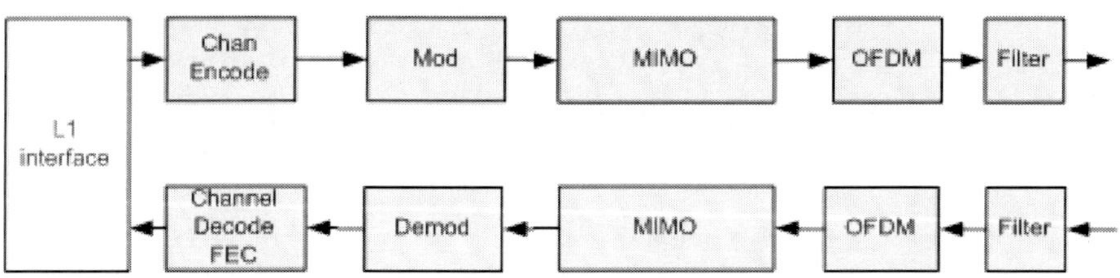

Figure 13. LTE SBS Main Baseband Processing Components.

Total power consumed by a BS station comprises digital baseband (BB), RF (analogue), power amplifier and overhead (power systems and cooling) [20]:

$$P_{total} = P_{\mathrm{BB}} + P_{\mathrm{RF}} + P_{\mathrm{PA}} + P_{\mathrm{overhead}} \tag{1.15}$$

For the purposes of indoor SBSs the power associated with cooling can be ignored. In addition, the generation of power e.g., AC-DC and DC-DC conversion is omitted from this analysis to be comparative with the 2D-Rake analysis presented earlier.

Using appropriate power and scaling tables given in [20], and based on the following supplied expression (1.16) the power consumed for a desired set of key baseband sub-components can be determined, namely: filter: up/down sampling and filtering, OFDM: FFT and OFDM-specific processing, Frequency domain (FD) processing, mapping, MIMO equalization, foreword error correction (FEC).

$$P_{\text{total}} = \sum_{i\in I_{BB}} P_{i,\text{ref}} \prod_{x\in X} \left(\frac{x_{act}}{x_{ref}}\right)^{s_{i,x}} + \sum_{i\in I_{RF}} P_{i,\text{ref}} \prod_{x\in X} \left(\frac{x_{act}}{x_{ref}}\right)^{s_{i,x}} + P_{\text{PA}} + P_{\text{Overhead}} \quad (1.16)$$

Considering the complexity of an LTE SBS: Assuming the power of the frequency domain linear processing of a 20 MHz, 4 x 4 MIMO, 16-QAM, coding rate 1, with a 100% of time-domain duty cycle and a 100% frequency occupation, the power can be computed from (1.15) as:

$$P_{\text{FD.lin}} = P_{\text{FD.lin,ref}} \left(\frac{BW_{\text{act}}}{BW_{\text{ref}}}\right)^{s_1} \left(\frac{M_{\text{act}}}{M_{\text{ref}}}\right)^{s_2} \left(\frac{R_{\text{act}}}{R_{\text{ref}}}\right)^{s_3} \left(\frac{Ant_{\text{act}}}{Ant_{\text{ref}}}\right)^{s_4} \left(\frac{dt_{\text{act}}}{dt_{\text{ref}}}\right)^{s_5} \left(\frac{df_{\text{act}}}{df_{\text{ref}}}\right)^{s_6} \quad (1.17)$$

where n_{act} and n_{ref} referred to the actual system component under scrutiny and the reference system (20 MHz, single antenna, 64-QAM, coding rate 1, 100% time domain and frequency domain duty cycling) respectively. **s** is the scaling vector (1 or 0). Where *BW* in the bandwidth, *M* is the modulation index, *R* is the FEC coding rate, *Ant* is the number of antennas, *dt* is the time-domain duty cycling and *df* is the frequency domain duty cycling.

Using the power required for DL linear processing ($P_{\text{FD.lin,ref}}$ = 0.166 W, with 120 GOPS/W in 65 nm technology), and assuming that the FBS/SBS employs highly efficient integration of signal processing with power efficient dedicated hardware, we have the following power estimation (1.18). This assumes a 10 MHz bandwidth, 64-QAM, 4 antennas, 100% time and frequency domain duty cycling and an FEC code rate of 1.

$$= 0.167\text{W} \times \left(\frac{10}{20}\right)^1 \left(\frac{6}{6}\right)^0 \left(\frac{1}{1}\right)^0$$

$$\left(\frac{4}{1}\right)^1 \left(\frac{100}{100}\right)^1 \left(\frac{100}{100}\right)^1 = 0.334\text{W} \quad (1.18)$$

Completing the analysis of the LTE SBS baseband based on the approach above, using (1.17) gives approximated total power attributed to the baseband as per Table 5.

Factoring in the transceiver into (1.17) and considering a range of values for b to be between 32 - 4096 mw, increases the overall power as Table 6. The Energy Consumption Ratio ECR (1.19) is given based on a theoretical bitrate of 480 Mbits/s where E is the energy required to deliver M bits of information over time T, and $D = M/T$ is the data rate in bits per second.

$$ECR = \frac{E}{M} = \frac{PT}{M} = \frac{P}{D} \; [J/bit] \quad (1.19)$$

Table 5. Baseband Power – LTE SBS (10 MHz, 64 QAM, 4 antennas, ½ rate Turbo)

UL/DL	Ref (GOPS)	Power (W), 65 nm, 2010	Power (W), 45 nm, 2010
UL			
Filter (inc D/C)	150	2.5	1.25
OFDM	60	1	0.5
FD Lin	30	0.5	0.25
FD non-lin	10	0.167	0.083
FEC	110	1.833	0.917
DL			
Filter	100	1.667	0.833
OFDM	60	1	0.5
FD lin	20	0.333	0.167
FD non-lin	5	0.333	0.167
FEC	20	0.333	0.167
Total Power (W)		9.667	4.833

4.1.1. 2D-Rake

The 2D-rake power calculated above, considered a power efficiency of the 12.8 GFLOPS/W. In order to make a comparison with the LTE system, the power efficiency of 120 and 240 GOPS/W should be used. Based on this, the results are shown in Table 7 and Table 8 for both Architecture 1 and Architecture 2. ECR is provided based on the link budget shown in Table 2 and assumes a 10 m cell at 28 GHz. The use of state of the art mmWave ADC/DAC components [22],[23] (which include the necessary up and down conversion), are included in the total power. A search rate of 180 Hz is assumed.

4.1.2. SE-EE Tradeoff 4G vs. 5G

Based on the overall circuit power consumption calculated above for the LTE FBS, 2D-Rake architecture 1 and 2D-Rake architecture 2, the SE-EE tradeoff is shown in Figure 14. Circuit powers assume worst case transceiver power for all architectures.

Table 6. Total Power – LTE (10 MHz, 64 QAM, 4 antennas, ½ rate) Turbo)

Transceiver power, *b*, (mw)	Total Power (W), 65 nm, 2010	ECR J/bit	Total Power (W), 45 nm, 2012	ECR J/bit
32	9.795	2.04E-8	4.961	1.03E-8
64	9.923	2.07E-8	5.089	1.06E-8
128	10.179	2.12E-8	5.345	1.11E-8
256	10.691	2.23E-8	5.857	1.22E-8
512	11.715	2.44E-8	6.881	1.43E-8
1024	13.763	2.87E-8	8.929	1.86E-8
2048	17.859	3.72E-8	13.025	2.71E-8
4096	26.051	5.42E-8	21.217	4.42E-8

Table 7. Baseband and Transceiver Power – Architecture 1

Transceiver power, *b*, (mw)	Arch 1 power (W), 65 nm, 2010	ECR J/bit	Arch 1 power (W), 45 nm, 2012	ECR J/bit
32	4.70	1.51E-9	4.38	1.41E-9
64	4.77	1.54E-9	4.45	1.43E-9
128	4.89	1.57E-9	4.57	1.47E-9
256	5.15	1.66E-9	4.83	1.56E-9
512	5.66	1.82E-9	5.34	1.72E-9
1024	6.69	2.16E-9	6.37	2.05E-9
2048	8.73	6.71E-9	8.41	2.71E-9
4096	12.83	4.13E-9	12.51	4.03E-9

Table 8. Baseband and Transceiver Power – Architecture 2

Transceiver power, *b*, (mw)	Arch 2 power (W), 65 nm, 2010	ECR J/bit	Arch 2 power (W), 45 nm, 2012	ECR J/bit
32	9.34	3.01E-9	8.7	2.80E-9
64	9.4	3.03E-9	8.77	2.33E-9
128	9.53	3.07E-9	8.89	2.87E-9
256	9.79	3.15E-9	9.15	2.95E-9
512	10.3	3.32E-9	9.66	3.12E-9
1024	11.32	3.65E-9	10.69	3.45E-9
2048	13.37	4.31E-9	12.73	4.11E-9
4096	17.47	5.63E-9	16.83	5.43E-9

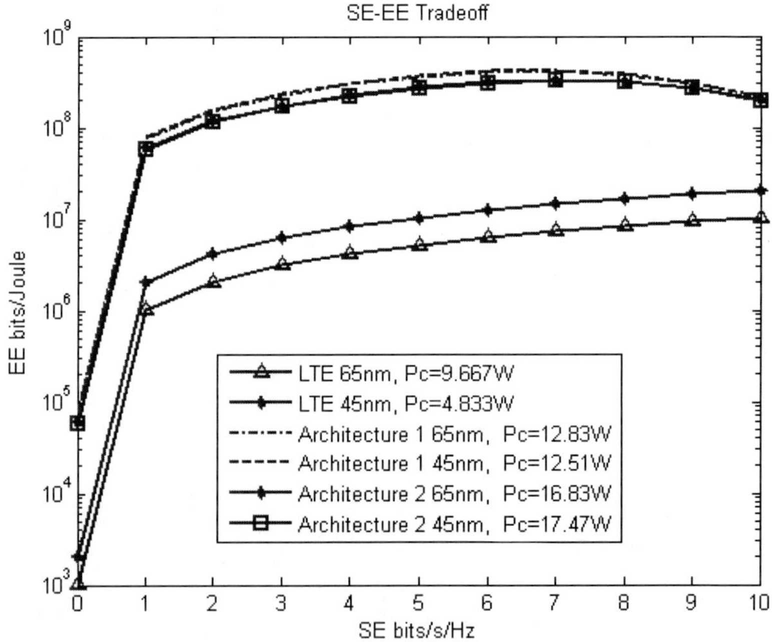

Figure 14. 4G LTE vs. Proposed 5G Architectures.

Based on these results, both 2-D Rake architectures are shown to provide superior performance and improve the energy consumption. Significantly, EE-SE tradeoff is shown to improve by an order of magnitude. The complexity analysis of the 2D-Rake architectures concentrated heavily on the baseband, however realistic ADC and DAC powers were included in the analysis. The use of RF beamforming means that the only a single ADC and DAC are used in the implementation.

Conclusion

This chapter has analysed the possible air interface approaches suitable for indoor small cell use. With energy efficiency being critical, the use of multiple antenna beamforming may be limited to a small number of antennas resulting in a wider beamwidth and therefore subject to multipath and ISI. To exploit this, the use of direct sequence spread spectrum using a Rake combiner is analysed. The Rake combiner used in conjunction with a moderate number of beamforming antennas is shown to give good theoretical performance as well as providing gain from the multipath. Two architectures were presented: one using a single chain RF beamformer and one using transmit and receive diversity. The complexity of the solutions was analysed in detail to provide an anticipated power usage based on the physical layer signal processing operation. The impact of multipath search resolution was shown to adversely impact power consumption, but based on an analysis of the search rate required in the small cell, dramatic reductions in total power were shown. Finally, the solutions were compared to that of present day 4G LTE technology for their power usage and energy efficiency. Results showed improved Energy Consumption Ratio for the two candidate architectures and provided a SE-EE tradeoff significantly better than a current generation 4G FBS by an order of magnitude.

References

[1] Da Silva, I.; Mildh, G.; Rune, J.; Wallentin, P.; Vikberg, J.; Schliwa-Bertling, P.; Rui Fan, "Tight Integration of New 5G Air Interface and LTE to Fulfill 5G Requirements," *Vehicular Technology Conference (VTC Spring), 2015 IEEE 81st*, vol., no., pp. 1,5, 11-14 May 2015.

[2] Dahlman, E.; Mildh, G.; Parkvall, S.; Peisa, J.; Sachs, J.; Selén, Y.; Sköld, J., "5G wireless access: requirements and realization," In: *Communications Magazine, IEEE*, vol. 52, no. 12, pp. 42-47, December 2014.

[3] Demestichas, P.; Georgakopoulos, A.; Karvounas, D.; Tsagkaris, K.; Stavroulaki, V.; Jianmin Lu; Chunshan Xiong; Jing Yao, "5G on the Horizon: Key Challenges for the Radio-Access Network," *Vehicular Technology Magazine, IEEE*, vol. 8, no. 3, pp. 47, 53, Sept. 2013.

[4] Larew, S.G.; Thomas, T.A.; Cudak, M.; Ghosh, A., "Air interface design and ray tracing study for 5G millimeter wave communications," In: *Globecom Workshops (GC Wkshps), 2013 IEEE*, vol., no., pp. 117-122, 9-13 Dec. 2013.

[5] T. Marzetta, "Noncooperative cellular wireless with unlimited numbers of base station antennas," Wireless Communications, *IEEE Transactions on*, vol. 9, no. 11, pp. 3590-3600, November 2010.

[6] Rowell, C.; Shuangfeng Han; Zhikun Xu; I Chih Lin, "Green RF technologies for 5G networks," In: *Wireless Symposium (IWS), 2014 IEEE International*, vol., no., pp. 1-4, 24-26 March 2014.

[7] Zhouyue Pi; Khan, F., "An introduction to millimeter-wave mobile broadband systems," *Communications Magazine, IEEE*, vol. 49, no. 6, pp. 101, 107, June 2011.

[8] K.AO-Cheng Huang, Zhaooheng Wang, "Beam Steering Antenna" In: *Millimeter Wave Communication Systems. IEEE/J.* Wiley and Sons, 2012.

[9] Rappaport, T.S.; Shu Sun; Mayzus, R.; Hang Zhao; Azar, Y.; Wang, K.; Wong, G.N.; Schulz, J.K.; Samimi, M.; Gutierrez, F., "Millimeter Wave Mobile Communications for 5G Cellular: It Will Work!," *Access, IEEE*, vol. 1, no., pp. 335, 349, 2013.

[10] S. Deng, C. J. Slezak, G. R. MacCartney, Jr., and T. S. Rappaport, "Small Wavelengths Big potential: Millimeter wave propagation measurements for 5G," *Microw. J.*, Nov. 2014.

[11] Anderson, C.R.; Rappaport, T.S., "In-building wideband partition loss measurements at 2.5 and 60 GHz," In: *Wireless Communications, IEEE Transactions on*, vol. 3, no. 3, pp. 922-928, May 2004.

[12] Manabe, T.; Miura, Y.; Ihara, T., "Effects of antenna directivity and polarization on indoor multipath propagation characteristics at 60 GHz," In: *Selected Areas in Communications, IEEE Journal on*, vol. 14, no. 3, pp. 441-448, Apr. 1996.

[13] Sawaya, W.; Clavier, L., "Simulation of DS-CDMA on the LOS multipath 60 GHz channel and performance with RAKE receiver," In: *Personal, Indoor and Mobile Radio Communications, 2003. PIMRC 2003. 14th IEEE Proceedings on*, vol. 2, no., pp. 1232-1236 vol. 2, 7-10 Sept. 2003.

[14] Rendevski, N.; Cassioli, D., "60 GHz UWB rake receivers in a realistic scenario for wireless home entertainment," In: *Communications (ICC), 2015 IEEE International Conference on*, vol., no., pp. 2744-2749, 8-12 June 2015.

[15] Andrew. J. Viterbi, "Modulation and demodulation of spread spectrum signals," *CDMA Principles of Spread Spectrum Communications* 3rd ed. Chichester, UK: Addison Wesley, 1996, pp. 77-121.

[16] Hsueh-Jyh Li; Ta-Yung Liu, "Comparison of beamforming techniques for W-CDMA communication systems," In: *Vehicular Technology, IEEE Transactions on*, vol. 52, no. 4, pp. 752-760, July 2003.

[17] Singh, J.; Ponnuru, S.; Madhow, U., "Multi-Gigabit communication: the ADC bottleneck[1]," *Ultra-Wideband, 2009. ICUWB 2009. IEEE International Conference on*, vol., no., pp. 22, 27, 9-11 Sept. 2009.

[18] Muirhead, D.; Imran, M.A.; Arshad, K., "Insights and Approaches for Low-Complexity 5G Small-Cell Base-Station Design for Indoor Dense Networks," In: *Access, IEEE*, vol. 3, no., pp. 1562-1572, 2015.

[19] Ghosh, A.; Thomas, T.A.; Cudak, M.C.; Ratasuk, R.; Moorut, P.; Vook, F.W.; Rappaport, T.S.; MacCartney, G.R.; Shu Sun; Shuai Nie, "Millimeter-Wave Enhanced Local Area Systems: A High-Data-Rate Approach for Future Wireless Networks," *Selected Areas in Communications, IEEE Journal on*, vol. 32, no. 6, pp. 1152, 1163, June 2014.

[20] Desset, C.; Debaillie, B.; Giannini, V.; Fehske, A.; Auer, G.; Holtkamp, H.; Wajda, W.; Sabella, D.; Richter, F.; Gonzalez, M.J.; Klessig, H.; Godor, I.; Olsson, M.; Imran, M.A.; Ambrosy, A.; Blume, O., "Flexible power modeling of LTE base stations," In: *Wireless Communications and Networking Conference (WCNC), 2012 IEEE*, vol., no., pp. 2858-2862, 1-4 April 2012.

[21] Hong Yang; Marzetta, T.L., "Total energy efficiency of cellular large scale antenna system multiple access mobile networks," In: *Online Conference on Green Communications (GreenCom), 2013 IEEE*, vol., no., pp. 27-32, 29-31 Oct. 2013.

[22] ADC12J4000 12-Bit 4 GSPS ADC With Integrated DDC12-Bit. http: //www.ti.com/lit/ds/symlink/adc12j4000.pdf.

[23] DAC39J84 Quad-Channel, 16-Bit, 2.8 GSPS, Digital-to-Analog Converter. http://www.ti.com/lit/ds/symlink/dac39j84.pdf.

In: Energy Efficiency
Editor: Julian E. Perez
ISBN:978-1-53611-040-1

Chapter 6

EVALUATING THE ENERGY EFFICIENCY OF VIRTUALISED 5G NETWORKS

Raad S. Alhumaima, Shireen R. Jawad and H. S. Al-Raweshidy*
Wireless Networks and Communications Centre (WNCC),
Department of Electronic and Computer Engineering,
College of Engineering, Design and Physical Sciences,
Brunel University London, Uxbridge, Middlesex, UK

Abstract

In virtualised 5G networks, evaluating the Power Consumption (PC) and other trade-offs could help optimising the power planning, quality of service (QoS) and Energy Efficiency (EE). In this chapter, parameterised and componentised power models (PMs) are proposed to evaluate the virtualisation based software defined cloud-radio access network (VSDC-RAN) paradigm. This evaluation requires a knowledge about the PC of each individual component which constitute the virtualised server. Cooling and total PC for this architecture with and without virtualisation have been compared using differentiated parameters, such as varying number of bare-metal Base Band Units (BBUs), VMs and system's bandwidth share. The results show dramatic decrease in the total PC via virtualising the core network (CN), whereas the trade-off of such gain is extensively demonstrated.

*E-mail address: 1234914@my.brunel.ac.uk

1. Introduction

Amongst all suggested 5G paradigms and technologies, the mobile operators and equipment vendors have privileged Cloud Radio Access Network (C-RAN) implementation [1], [2]. C-RAN architecture included the base band units (BBUs) that are hosted in a BBU pool/cloud, and remote radio heads (RRHs). The BBU is responsible for processing all higher and most of the physical layers functions. In turn, the RRHs are left in charge of the radio frequency simple functions. By doing so, C-RAN curtails both the operational and capital expenditures (OPEX) and (CAPEX) respectively due to reduced maintenance and less site visits and leases cost. Additionally, C-RAN enables BBUs cooperation, dynamic spectrum utilisation, reduction of cooling and total power consumption (PC). As C-RAN intensifies the number of RRHs, the control plane functionality in the core network (CN) is intensified too, this is generally related to users (UEs) call sessions' set up, handover and authentications. Thereafter, enhancing the scalability of the network by proposing innovative techniques becomes a precondition. Recently, the research community has embraced the introduction of network function virtualization (NFV) and software defined networks (SDNs) as paramount for the cloud architecture problems [3].

The latter offers a centralised and unified network administration by assembling the control plane functions of the network devices to a dedicated controller(s). The controller masters the open flow (O.F) switches to authorise evaluating/assessing the network traffic. The controller is also responsible for maintaining the software abstracts, set up the application programming interfaces (APIs) to the other network entities/providers, and reserves an open protocol to update the underlying switches. On the other hand, the controller interfaces the legacy control plane units such as packets gateway (PGW), mobility management entity (MME) and serving gateway (SGW) via dedicated APIs to update and exchange UEs' admissions and authorisation operations. In [4], the complete signalling procedure and the functional design of SDN integrating the control plane units have been adequately described. Despite such planning (i.e., deploying the controllers and switches) can unleash the network potentials to the maximum. However, this imposes speculations about the efficiency of these networks in terms of PC. The added devices will burden the network with extra

PC, which might not be justified by the acquired network flexibility and programmability.

1.1 Foreseen Solution via NFV

NFV have recently been devoted to provide significant increase in the energy efficiency (EE). NFV allows service providers to execute network's functions using software rather than running proprietary built or dedicated appliances. Due to the latter, updating and expressing new services and applications is increasingly intractable. NFV therefore enables the use of general off-the-shelf equipment to run the functions' software, also known as virtual machines (VMs). However, running multiple VMs on a single hardware requires a VMs supervisor or manager. This manager is also called a HyperVisor (HV). The HV is a software that runs on the host's hardware to dynamically control and allow the host server to be shared by several VMs. Each VM then appears to utilise the hardware's RAM, processors (CPUs), Network Interface Cards (NICs) and HarD Drive (HDD) all to itself. The HV assures that the guest VMs cannot obstruct each others while accessing the server's resources. However, the presence of the HV in the host server increases the RAM accesses, CPU's functional processes and HDD's usage as a consequence of the increased VMs interruptions and orchestration. The HV therefore exhibits a measurable extra overhead added to the server. Such matter urges to evaluate this increment in the PC and other consequences/trade-offs within each host compared to the non virtualised servers. Furthermore, it is necessary to compare the entire PC of the CN that is reduced by using Virtualisation technology with the non virtualised CN [5].

1.2 Impacts of SDN and NFV up on C-RAN

1. SDN decouples the control and data plane to simplify the control plane design in the CN. This can reduce the CN entities's functional complexities and their overhead. However, NFV appears as a complementary to SDN by achieving more provisioning for the available servers' resources.

2. The delay produced by the signalling amongst the SDN controller and BBUs can be remarkably reduced. This is because the controller connects near located BBUs in the BBU pool rather than distributed BBUs. This

eventually means the legacy X-2 protocol signalling cost in the BBU pool is fully or partially relieved.

3. SDN deploys O.F switches nearby/within the RRHs, this import CN's control plane contents closer to the UEs. The installed flow tables of the switches permit straight communications with the UEs. Before approaching the CN, the packets can be directed to their final destination. As a result, the latency constraint can be further relaxed.

4. The relieved latency by SDN can be augmented by using NFV. As many VMs are encompassed in one host server, the HV accordingly enables each VMs to access the share of allocated resources contained by the host hardware in a time restricted manner. This process yields a slower response time when executing applications of virtualised as compared to a non virtualised servers [6]. However, optimising the number of the VMs installed in one host server is prerequisite in regards to the traffic demands and available server resources.

5. SDN is essentially proposed for wired rather than wireless networks. Accordingly, bandwidth optimisation and power allocation methods will be required in order to gain its operational execution.

6. The ability of migrating the VMs amongst the host servers makes possible to provide services form BBU pool locations that are optimal for present UEs' locations, this enables the VMs to act as latency enhancers [7].

Note that, PC probably indicates Power Consumed or power Consumption. We also refer to SDN and C-RAN integration as SDC-RAN, whereas integrating NFV and SDC-RAN as VSDC-RAN.

1.3 Literature Review

There have been several researches seeking to investigate modelling the network's PC on components and system level. Herein, merely the issue of system level is deliberated upon as it is the focal point within this study. In [8] and [9], the State of the Art (SotA) LTE base station's (BS) power models (PMs) are evaluated; these PMs can be characterized as fairly complex, seeing that they undergo the functional computations and the complexity of determining the PC of each network's component, so as the total PC can be estimated. These two models are simplified and parameterised by Energy Aware Radio and network TecHnologies (EARTH) PMs in [10] and [11]. The latter have been used to determine the SotA BSs' PMs by linearising the PC model of the BSs (i.e., Macro, Micro, Pico, etc.). However, these have been unable to estimate the amount of PC in the futuristic and hybrid networks, such as C-RAN, Heterogeneous C-RAN (H-CRAN) or SDN based C-RAN. Furthermore, two of our former studies conducted in regards to modelling the PC of the new and hybrid 5G networks can be referred to in [12] and [13]. The first study pursued investigation concerning the level of power reduction that can be attained if C-RAN is to be adapted instead of the traditional BSs deployment. In the second study, SDN has been integrated with C-RAN and such cooperation is evaluated concerning the PC. Additionally, compensating the increased PC due to SDN controller and O.F switches is discussed. The work also suggested to use NFV as a solution for this growth in the PC. There are several works mentioned in [14], presenting proposals, algorithms and paradigms to adapt and implement NFV and SDN in the cloud, for example: NeFuCloud, PLayer, CloudNaaS, APLOMB, PACE, SIMPLE, CloudNFV and REALTIME CLOUD. These researches investigate and define different mechanisms associated to the various modes of operations, and mostly they rely on open source tools to implement the suggested paradigms. Although, unfortunately these research studies lack mathematical representation, in-depth analysis and evaluation on the matter of power cost and allocations. In [5], the virtualisation effect upon the PC of a single server while running certain packages and applications is tested. This work only contemplate in association to a single server case study without providing a mathematical model based platform to measure the PC of server or system level, similarly in [15], [16] and [17].

1.4 Chapter Contributions

In large-scale computing centres, defining the way a server consumes power resources is an essential element of examination when developing power management approaches. In addition, the VM's PC has scarcely been evaluated during performing network tasks. Hardware Virtualization adds substantial overhead as a busy web-server consumes about 40% more power, and the VM may take approximately five additional cycles to transmit a packet with less efficiency, when compared to a non-virtualized server [5]. Therefore, to answer whether or not +rthwhile to compromise the network performance with PC reduction, the following contributions have been accomplished:

1. Modelling the way active VMs increasingly affect the PC of the host server in terms of CPU, RAM, NIC and HDD. Furthermore, the HV's PC that controls the mentioned resources is also modelled. This will provide a realistic visualisation to the PC modelling on device level.
2. Measuring the components' PM of VSDC-RAN considering only the CN units. This confers a component level based PC evaluation.
3. By folding the computations complexity of the components PM, a parameterised PM has been produced. This simplified model diminishes the manifold representation of the components model. But it also provide an accurate method to evaluate the networks PC.
4. Comparing total and cooling PC of SDC-RAN and VSDC-RAN by taking different parameters into consideration, such as number of VMs, BBUs and bandwidth used.

2. Server Power Consumption

As a consequence to virtualisation, each host server could experience increment in the PC due to higher computation levels, generated I/O instructions executions and compound accessing for the device resources by the aggregated VMs' applications [6]. Generally, there are four major participants involved within the constituency of a server PC: RAM, CPU, NIC and storage or HDD, the virtualised PM of each part can be expressed as follows:

2.1 RAM Power Model

Usually, each VM requires a share from the RAM to utilise during operation. However, when the number of installed VMs in one server (N) increases, the size of the RAM as well as its corresponding PC is also intensified in order to handle an amplified amount of RAM usage requests [18]. Therefore, it is compulsory to identify the appropriate RAM size (Z_{RAM}) to place in a server that contains N VMs. The proposed method initially describes the change in the RAM size corresponding to the change in N, i.e., $\frac{dZ_{RAM}}{dN} = \alpha Z_{RAM}$, when solving this equation, it yields $Z_{RAM}(N) = Z_{int}e^{\alpha N}$, where Z_{int} is the initial RAM size. Afterwards, the corresponding PC $P_{RAM}(Z_{RAM})$ of the RAM can be scrutinizes. Continuously, $P_{RAM}(Z_{RAM})$ can be modelled based on recognizing the maximum and initial PC of the RAM size. In this case, the constant ($\beta << 0.1$) is introduced to shape such change in the PC:

$$\frac{dP_{RAM}}{dZ_{RAM}} = \beta P_{RAM} \tag{1}$$

When solving equation (1), it yields:

$$P_{RAM}(Z_{RAM}) = P_{intRAM}\ e^{\beta Z_{RAM}} \tag{2}$$

Where P_{intRAM} is the initial PC of the RAM. It is worth noting that the constants α and β may vary based on type of the RAM used and its initial and maximum PC. However, bearing in mind, the above relation is straightforward when outlining how to increase Z while increasing N, nevertheless it does not estimate how much the RAM itself over-consumed in terms of power whilst increasing N. To consider this, Z is presented as $Z = 2^{i+1}GB$, where i is an index ($i = 0, 1, 2, 3...$) to shape the commercial RAM sizes i.e., $Z = (2, 4, 8, 16, 32GB, ...)$. After, if P^i_{RAM} is the PC of the RAM holds i index, then the change in PC of this specific RAM dP^i_{RAM} is correlated to the change of N (dN). To estimate this change, the constant ν is presented as follows:

$$\frac{dP^i_{RAM}}{dN} = \nu P^i_{RAM} \tag{3}$$

When solving equation (3), it produces:

$$P^i_{RAM}(N) = P^i_{intram}\ e^{\nu N} \tag{4}$$

Where P^i_{intram} is the initial PC of RAM size with index i. The RAM PC model, in Watt (W), can be finalised as follows:

$$P^i_{RAM}(Z_{RAM}, N) = (P_{intRAM} + \Delta)\ e^{\beta Z_{RAM}} + e^{\nu N} \tag{5}$$

Where $\Delta = P^{i}_{intram} - P_{intRAM}$, represents the power difference between the tagged initial RAM i PC and the least RAM size initial PC (P_{intRAM}).

2.2 CPU Power Model

It was assumed that the number of cores per CPU ($C \geq NC+1$), where NC is the number of VMs installed in one core. If K denotes the total number of CPUs per server, then $N = \sum_{k=1}^{K} \sum_{c=1}^{C} NC_{k,c}$.

Furthermore, the more VMs installed, the greater amount of power the core will consume until it reaches its maximum. Such affiliation is recognised as being exponential in a core PC level. Logically, the core PC (P_{core}) upsurges from an initial when increasing NC. The above assumptions can be translated into the following model:

$$\frac{dP_{core}}{dNC} = -\mu P_{core} \tag{6}$$

Where ($\mu < 0.1$) is a positive constant used to determine how the power is increased. Such a linkage tends to be linear when μ approaches 0. However, when solving equation (6), it yields as:

$$P_{core}(NC) = P_{intcore}\, e^{-\mu NC} \tag{7}$$

Where $P_{intcore}$ is the initial core PC. If P_{core} is calculated, the CPU's PC (P_{CPU}) can also be known by gathering the PCs of all cores:

$$P_{CPU} = \sum_{c=1}^{C} P_{core(c)} \tag{8}$$

Additionally, the total PC, in W, of CPUs per server (P^{CPU}_{server}) can be obtained as the following:

$$P^{CPU}_{server} = \sum_{k=1}^{K} P_{CPU(k)} \tag{9}$$

2.3 NIC Power Model

Each NIC is shared amongst multiple VMs simultaneously, and the higher amount of VMs found per server, the more NICs are required to serve them. Moreover, the PC of each NIC is obligated to an augmentation in the PC [19]. The maximum number of NICs (L) that can be placed in one server is shared amongst all the VMs (N), i.e., ($L \propto N$) where ($L \leq N$). For practical consideration, L is assumed equivalent to 8. Each NIC (l) then is assigned (N/L) VMs. Moreover, the PC of each NIC rises when the N/L is increased, as more packets will be received and transmitted via a particular NIC. Hence for that reason, the model has to outline an increment in the PC, so as the virtualised NIC is driven to reach its maximum PC, up from an initial value (P_{intnic}). If P_{nic} is the PC of the virtualised NIC, hence, the PC of NIC can be expressed to be ($P_{nic}(\frac{N}{L})$), as a function of (N/L).

This linkage refers to the upsurge of P_{nic} in correspondence to the change of $\frac{N}{L}$, therefore, such behaviour can be modelled as the following:

$$\frac{dP_{nic}}{d\frac{N}{L}} = \gamma P_{nic} \tag{10}$$

When solving (10), the following solution can be obtained:

$$P_{nic}(\frac{N}{L}) = P_{intnic} e^{\gamma \frac{N}{L}} \tag{11}$$

Next, the total PC, in W, of all NICs per server (P^{NIC}_{server}) can be calculated:

$$P^{NIC}_{server} = \sum_{l}^{L} P_{nic(l)} \tag{12}$$

2.4 Storage (HDD) Power Model

As a matter of fact, over-utilising the storage by installing more VMs increases the required cooling and total PC [20]. Therefore, the model has to contemplate two main behaviours of the storage: firstly, the storage is shared amongst the other servers in the CN. Secondly, the VMs per server drive the storage PC up, as they increase its data accesses. Subsequently, if the PC of the storage device is $P_{storage}$, hence, there will be two influencing parameters: the time at which the storage is being utilised (t) and N, i.e., $P_{storage}(t,N)$. As far as the first variable is concerned, it is considered that other servers can add, access and delete data from the tagged server. In this case, it is assumed that the storage capacity varies during time (t) to represent the sharing capability amongst the CNs' servers. If $P_{storage}$ is increased/decreased by time (t), it means adding/deleting supplementary contents to/from the storage by time (t). The second variable in turn clarifies how the added N VMs increase the storage PC, as a result of increasing the rate of accessing the stored data. When considering the time, two scenarios can be determined: the first is when the storage PC ($P^i_{storage}$) increases by time, following the exponential method, the PC can be formulated as:

$$\frac{dP^i_{storage}}{dt} = -\delta P^i_{storage} \tag{13}$$

Where ($\delta < 1$) is a positive constant, which can control the maximum value of the storage PC, when solving (13), it yields:

$$P^i_{storage}(t) = P_o e^{-\delta t} \tag{14}$$

Where P_o is the initial storage PC. The second case is when the storage PC ($P^d_{storage}(t)$) decreases by time, following the same procedure as in (13), the PM can be expressed as follows:

$$P^d_{storage}(t) = 2P_o - P_o e^{-\delta t} = P_o[2 - e^{-\delta t}] \tag{15}$$

The amount ($2P_o$) is added to uplift the start of the curve ($P^d_{storage}(t)$) to start from the initial PC (P_o) at time ($t = 0$), so as $P^d_{storage}(t)$ and $P^i_{storage}(t)$ start from the same point.

The second variable of the storage is modelled when the PC is proportionally increased with the number of VMs (N). The procedure of (13) is followed and the resulting PM can be obtained as follows:

$$P_{storage}(N) = P_o e^{\xi N} \tag{16}$$

Where ξ is a positive constant. Note that the lesser value of the constants ξ and δ are assigned, the more the model approaches to be linear.

As the effect of the two variables t and N is independent of each other, the model separates their PC, yet the model of each variable is aggregated, i.e.,

$P_{storage}(t)$ will be added to $P_{storage}(N)$. The total storage PC, in W, of the virtualised storage then can be finalised:

$$P_{storage}(t,N) = \begin{cases} P_o(e^{-\delta t} + e^{\xi N}) & \text{if } \delta \text{ increases} \\ P_o(2 - e^{-\delta t} + e^{\xi N}) & \text{if } \delta \text{ decreases} \end{cases} \quad (17)$$

2.5 HyperVisor (HV) Power model

It was assumed that the number of HV's assignment tasks is *AS*, and the PC per task per VM is *PAS*, then the PC of the HV can be modelled as follows:

$$P_{HV} = \sum_{n=1}^{N} \sum_{as=1}^{AS} PAS_{(as,n)} \quad (18)$$

Where $PAS_{(as,n)}$ is the PC of the as-th job allocated to n-th VM .

Finally, the total PC, in W, of a server (P_{srvr}) is equivalent to the PCs of all components:

$$P_{srvr} = P_{RAM} + P^{NIC}_{server} + P^{CPU}_{server} + P_{storage} + P_{HV} \quad (19)$$

The previous description presents how the number of VMs affect server's PC. Herein, the modelling can go further to elaborate a single VM PC and its operative factors. The degraded performance can be modelled regarding the increased processed RBs within each VM by time. The increased dynamic and bandwidth based load is expressed by the number of RBs processed at each VM. This number increases the PC of this particular VM linearly or convexly. This is constrained by the PC summation of all VMs sharing a particular server component cannot exceed its maximum PC when the RBs are highly utilised. The higher processed RBs at each VM, the more share from the finite server resources/units is demanded and this eventually decline the performance of each VM and subsequently the entire network. The total number of VMs itself cannot be understood unless the load/RBs of each VM is illustrated. Therefore, the total number of processed RBs ($RB_T = \sum_n^N RB_n$) is added to N based description. In this model, the processed RBs is concerned to add extra overhead. This overhead is added to the previous modelling to produce the total PC of the virtualised BBU server (P_{server}), as follows:

$$P_{server} = P_{srvr} + e^{\vartheta * RB_T} \tag{20}$$

where ϑ is the increment factor due to processing RB_T by all VMs in any of the server resources. RB_T can reach up to 900 RBs to be executed at a time in some of the advanced servers. However, ϑ can be adjusted so as the total server PC cannot exceed its maximum.

This modelling of server PC is very simple but accurate, as it is mainly based on easy to obtain parameters, these are RBs, number of VMs, device initials and efficiencies from components data sheets. This modelling is different from what the research have provided, which are classified as intrusive, machine learning and software based models according to the approach used to measure the PC. Generally, these models are expected to add additional PC to the measured PC. Intrusion based models require to install intrusive tools, events counters which makes the PC measuring expensive and complex. Software based models require additional application or separate device to run, this method is also power consuming and complex. Machine learning algorithms are based on heuristics, this method however is time and power consuming. These algorithms may require additional device to run on.

3. Components Power Model

Figure 1 demonstrates the participating PC modules of VSDC-RAN, which encompasses mainly three parts:

- Virtual BBU pool (vBBU pool).
- Virtual control plane units (i.e., vMME, vSGW and vPGW).
- Virtual SDN controller and switch (vSDN controller and vSDN switch).

Noticing that, the terms (virtual XX) or (vXX) indicates that the server (XX) is undergoing a virtualisation technology. Furthermore, the PC of vCN is described as follows:

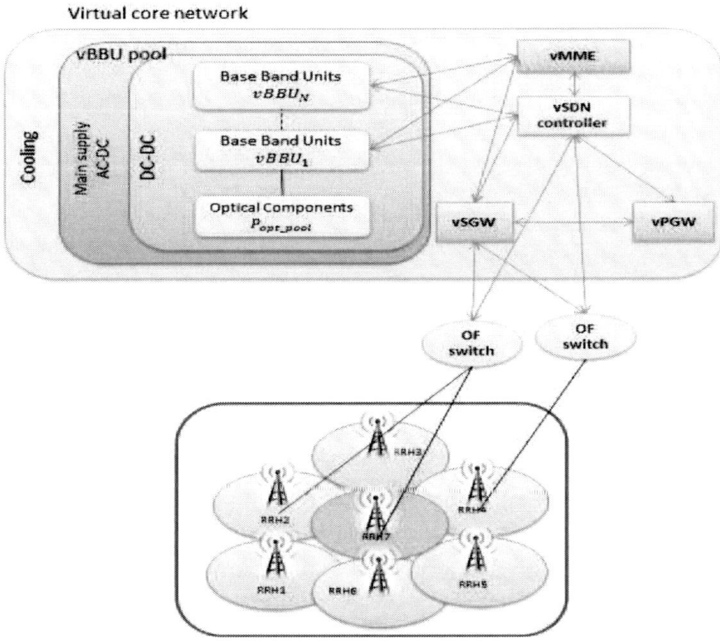

Figure 1. PC paradigm of VSDC-RAN.

3.1 Virtual BBU pool PC

The detailed BBU core PC (P_{core}^{BBU}) is described in [13], has been used to complete the vBBU server PC, in W, as follows:

$$P_{server}^{BBU} = \sum_{s1=1}^{S1} [P_{RAM} + P_{server}^{NIC} + P_{BBU}^{CPU} + P_{storage} + P_{HV} + e^{\vartheta * RB_T}]_{s1} \quad (21)$$

Where $S1$, is the total number of servers that used to host the BBUs' VMs. Where

$$P_{BBU}^{CPU} = \sum_{k1=1}^{K1} \sum_{c1=1}^{C1} P_{core(k1,c1)}^{BBU} \quad (22)$$

Furthermore, the CPU PC P_{MME}^{CPU}, P_{SGW}^{CPU} and P_{PGW}^{CPU} of MME, SGW and PGW, accordingly, is equivalent to the sum of their corresponding cores' PC:

$$P_{MME}^{CPU} = \sum_{k2=1}^{K2} \sum_{c2=1}^{C2} P_{core(c2)}^{MME} \tag{23}$$

$$P_{SGW}^{CPU} = \sum_{k3=1}^{K3} \sum_{c3=1}^{C3} P_{core(c3)}^{SGW} \tag{24}$$

$$P_{PGW}^{CPU} = \sum_{k4=1}^{K4} \sum_{c4=1}^{C4} P_{core(c4)}^{PGW} \tag{25}$$

P_{core}^{MME}, P_{core}^{SGW} and P_{core}^{PGW} denote the core PC of MME, SGW and PGW, which is responsible of MME, SGW and PGW functions, respectively. Comparably to BBU functions, PGW, MME, and SGW cores PC can be modelled following the same style, by taking into consideration the functions set of each unit [13]. $C1$, $C2$, $C3$ and $C4$ denote the number of BBUs, MMEs, SGWs and PGWs cores, correspondingly. $K1$, $K2$, $K3$ and $K4$ are the number of BBUs, MMEs, SGWs and PGWs' CPUs, respectively. Next, the PC of control plane server is introduced as:

$$P_{server}^{cl} = \sum_{s2=1}^{S2} [P_{RAM} + P_{NIC} + P_{cl}^{CPU} + P_{storage} + P_{HV} + e^{\vartheta * RB_T}]_{s2} \tag{26}$$

Where $S2$ denotes the total number of servers that host the control plane's VMs. P_{CPU}^{cl} in turn can be modelled as

$$P_{CPU}^{cl} = P_{CPU}^{MME} + P_{CPU}^{SGW} + P_{CPU}^{PGW} \tag{27}$$

3.2 Virtual SDN PC

SDN PC can be divided into two main parts: switches and controllers PCs:

3.2.1 O.F switch PC

If it is assumed that the switch is receiving/transmitting a number of packets. These packets will be matched with the installed flow tables, and accordingly actioned and directed to selected destinations. However, the CPU PC, in W, of the switches P_{switch}^{CPU} can be modelled:

$$P^{CPU}_{switch} = \sum_{k5=1}^{K5} \sum_{c5=1}^{C5} P^{switch}_{core(k5,c5)} \tag{28}$$

Where P^{switch}_{core} denotes the switch PC. $C5$ and $K5$ is the total number of cores and CPUs, respectively, that operate as O.F switches. In addition, the switches servers PC P^{switch}_{server} can be modelled as:

$$P^{switch}_{server} = \sum_{s3=1}^{S3} [P_{RAM} + P_{NIC} + P^{CPU}_{switch} + P_{storage} + P_{HV} + e^{\vartheta * RB_T}]_{s3} \tag{29}$$

Where $S3$ denotes the total number of servers that contains O.F switches' functions. If the switches are deployed as individual devices and their functions are implemented in separate switch devices rather than virtual functions. Then the switches server's PC above is ignored and the switch device PC can be jointly added to the total PC.

3.2.2 SDN controller PC

Since the SDN controller interfaces the other components in the CN to exchange signalling information. Therefore, the controller PM is based on the rates of outgoing control plane signals that are received by the switchs, PGWs, MMEs and SGWs from the SDN controller, in addition to the set of energies required for sending or receiving these packets. The CPU controller PC P^{CPU}_{SDNctl}, in W, can be evaluated as follows:

$$P^{CPU}_{SDNctl} = \sum_{k6=1}^{K6} \sum_{c6=1}^{C6} P^{SDNctl}_{core(k6,c6)} \tag{30}$$

P^{SDNctl}_{core} is the core PC of the controller, while the server that runs the controller's functions can be expressed as:

$$P^{SDNctl}_{server} = \sum_{s4=1}^{S4} [P_{RAM} + P_{NIC} + P^{CPU}_{SDNctl} + P_{storage} + e^{\vartheta * RB_T}]_{s4} \tag{31}$$

Where $S4$ denotes the total number of servers that hold the controllers functions. Furthermore, the detailed P^{switch}_{core} and P^{SDNctl}_{core} can be found in [13].

3.3 Overhead PC

DC-DC PC (P_{DC}), Mains supply PC (P_{MS}) and cooling PC (P_{cool}) can be expressed as a losses, their PC considerations can be found in [13].

4. Parameterised Power Model

Moving forward from the assumptions previously identified in Sections 2 and 3, it is of benefit to further simplify the complexity of the functions correlated with each component by using a simplified or parameterised index. Moreover, it is important to include the traffic variations and the effects of losses in a straightforward manner. This is done to offer an easy but accurate way to calculate the network PC without undergoing the computations of each component. In order to achieve this, the following analysis and approximations can be carried out:

- The CN's units scale linearly with the bandwidth share (B). This will show the effect of the increased traffic/UEs connected to the network.
- The AC-DC, DC-DC and cooling PC are linearly scaled with other CN's components PC. Furthermore, these units' effect up on the PM is approximated using loss factors (σ_{DC}, σ_{MS}, σ_{cool}) to represent DC, AC, cooling loss factors, respectively.
- The backhaul connections in the CN amongst the whole units is expressed as loss factor ($\sigma_{backhaul}$), which linearly scaled with the total number of CN units.

Successively, the maximum PC ($P_{VSDC-RAN}$), in W, is formulated as follows:

$$P_{VSDC-RAN} = \frac{B.(P^{BBU}_{server} + P^{cl}_{server} + P^{SDNctl}_{server} + P^{switch}_{server} + \sum_{r=1}^{R} P_{opt_r})}{(1-\sigma_{DC})(1-\sigma_{MS})(1-\sigma_{cool})(1-\sigma_{backhaul})} \quad (32)$$

The main objective of this simplification is to provide a general platform for the network providers to measure their entire network consumption with the possibility to skip over the complexity of the components PM.

5. Results

The presented model has been used to evaluate the consumed power of VSDC-RAN in terms of bandwidth and varying number of BBUs, SDN devices and control plane units. To correlate the findings of this chapter with a real time measurement, some of the resulting parameters were selected from [10] when possible, which represent the SotA PC measurements, these parameters are shown in Table 1. Furthermore, it was noticed that the PC of an Intel Server System type SR1690WB is 76.89W, which includes Intel Xeon E5620 processor (4 cores) [21]. However, to obtain the PC of each unit in the server, this value is divided amongst CPU, RAM and HDD, which gross 59.6W. The CPU is assumed to consume 29.6W, RAM is 4W, NIC is 2W and HDD is 25W, while the rest of the power was consumed by the overhead. Besides, the other components which are not affected by the number of VMs N such as Motherboard, fans, etc., are not included in the results.

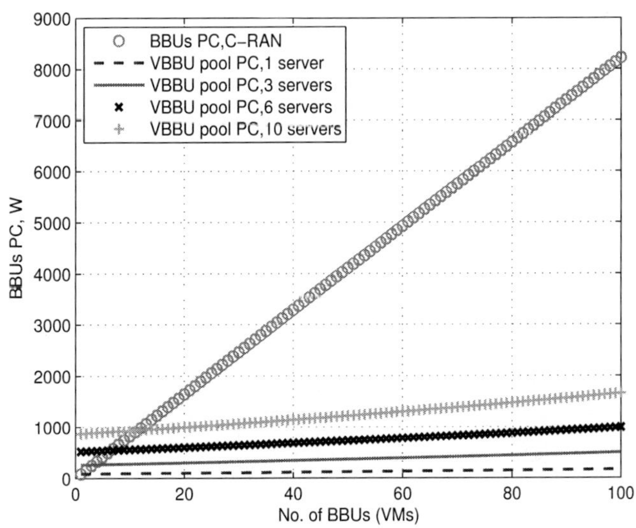

Figure 2. Comparison of the BBU pool PC with and without virtualisation of 100 BBUs/VMs

Figure 2 shows a comparison of BBU pool's PC with and without virtualisation. The maximum number of VMs software instances representing the BBUs is equivalent to 100 with different number of host servers (i.e., 10, 6, 3 and 1 servers). By using the rule of percentage change ($\frac{V1-V2}{|V1|} * 100\%$), it is found that the reductions in the PC are (78, 87.8, 93.9 and 97.5 %), respectively. These values are obtained when four virtualised cases representing only the BBU pool, are compared with the non virtualised servers.

On the same basis, Figure 3 shows a comparison of the entire CN's PC with and without virtualisation. There assumed 1 server hosts 2 VMs, these are a controller and switch VMs, while another server is used to host 3 control plane VMs (MME, SGW and PGW). Furthermore, for BBU pool, two scenarios were reflected, the first is when 5 servers were used in the BBU pool. In this case, a 82.3% reduction in the total PC can be achieved. The second is when 15 servers were used, this has produced 62% total PC reduction compared to the non virtualised CN.

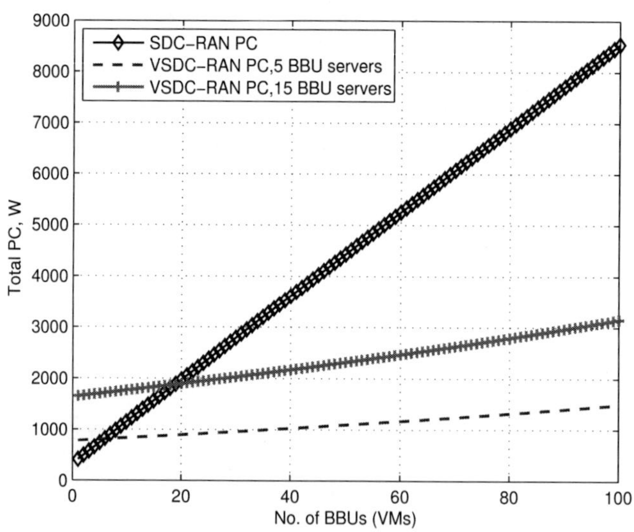

Figure 3. Comparison the total PC of the CN with and without virtualisation of 100 BBUs/VMs.

Continuously, Figure 4 shows cooling PC comparison of the CN. There are three cases that were compared with the non virtualised case, i.e., 3, 6 and 10 virtualised servers to serve the maximum number of VMs. If the percentage change is compared amongst the results of Figure 4, it yields (94.1, 88.2 and 80 %) PC reduction.

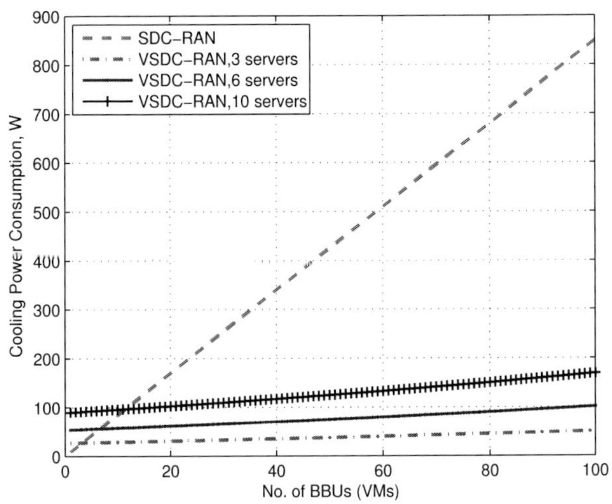

Figure 4. Comparison of cooling PC of the CN with and without virtualisation.

It is worth noting that if another types of servers hold different specifications are used in this comparison, the eventual results of these comparisons will also change. This is because such change instantly affects the initial and maximum consumption of each component within the virtualised servers, which slightly affects the outcomes. On the other hand, the more number of bare servers are considered in the comparison, the more PC gain. The reason is that any large number of servers will be multiplied by the entire server's PC, while in virtualisation method, this large number will only be multiplied by N, which only affects the host server's resources (CPU, RAM, etc.). Emphasising on the context of showing how the VMs increasingly affect the PC, only $P^i_{storage}(t)$ of the storage PC is considered. Furthermore, only the exponential RAM PM is adapted in the results.

Table 1. Used Parameters

Component	Unit	Value
P^{pm}_{switch}	W	58
$P^{pm}_{SDNctrl}$	W	20
P_{opt}	W	1
P^{BBU}_{server}	W	29.6
P^{cl}_{server}	W	29.6
P^{SDN}_{server}	W	29.6
P_{intRAM}	W	4
$P_{intcore}$	W	3.7
P_{intnic}	W	2
P_{o}	W	10
σ_{DC}	-	0.075
σ_{MS}	-	0.09
$\sigma_{backhaul}$	-	0.085
σ_{cool}	-	0.1
β	-	0.003
μ	-	0.1
γ	-	0.08
δ	-	-0.001
ξ	-	0.007
ν	-	0.005

If the bandwidth usually describes the traffic demands within the servers' components. Figure 5 shows the allocated bandwidth usage amongst the dedicated servers. It was assumed that the CN contains 70 servers, then this construction was compared with the virtual CN. There are three cases considered for the latter: 3, 6 and 10 virtualised host servers.

Although the HV PC is modelled separately than the other server's components (Section 2.5), its PC considerations are embedded within the model and not jointly added to the total PC value. Once the HV shares the host server's electronics, these chips however must not exceed their maximum power. Accordingly, if the HV PC is separately added to the server's RAM, CPU, etc., this

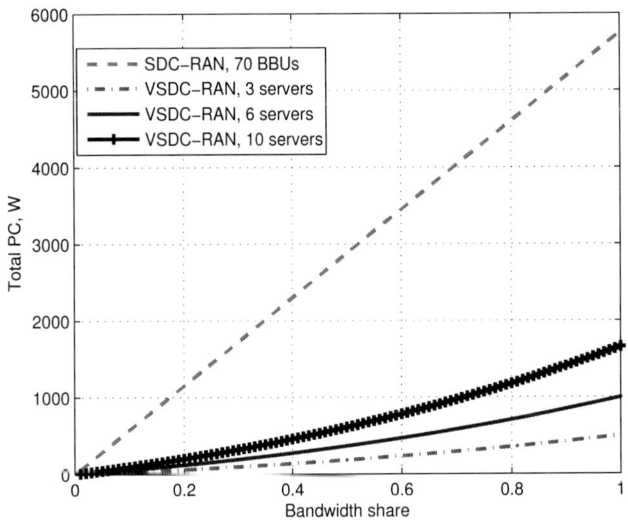

Figure 5. SDC-RAN's PC comparison with and without virtualisation as a function of bandwidth sweep.

will overtake the real assumptions and the physical realisation regarding the PC of host server's components. This conduct is already considered while modelling each component's PM. Nevertheless, the HV can be considered as one of the VMs that contributes to the PC calculations and included within N.

In terms of accuracy probing, the model mainly relies on the manufacturer specifications and design of each component (i.e., components' data sheets). As each equipment holds different operating conditions such as: type, initial and maximum PC, cooling requirement and efficiency, the outcome of the PM will be accordingly affected. On account of such variation is required to be adjusted through the constants mentioned in Sections 2 and 3 (i.e., α, β, etc.), the power tolerance of each component can be precisely found. Accordingly, a matrix of comparisons can be emerged by changing each constant's value in the above mentioned Sections. However, such assessments provides flexibility in the PM and a general platform viable for all the network service providers.

Conclusion

A components and parameters based PMs have been presented to demonstrate the PC calculations of the architecture VSDC-RAN. The effect of virtualisation is presented by providing a comparison for cooling and total PC before and after applying NFV. This comparison was subjected to different parameters such as varying number of BBUs, VMs and bandwidth sweep.

Intuitively, integrating SDN equipment with C-RAN upgrades the PC. However, the benefits and characteristics gained by SDN cannot be fully utilised unless NFV technology compensate such increment. Accordingly, by using the proposed PM, the power cost and gain of NFV and SDN can be assessed, the model also enables the network providers to distinguish the PC reduction while bearing the predicaments of NFV. The latter in turn is a time-critical service, which may fail to provide the same performance as the traditional servers. To justify such issue, there have been several works proposed recently to relief the problems of the HVs scheduling procedure, such problems are synchronization, real-time constraints, security, VMs placement and performance. These works however are proposed to unleash and extend the HV capability to an optimum, for example [22] and [23]. The investigations also promise new futuristic techniques, protocols, algorithms and designs to be innovated. Accordingly, NFV has been advocated to be used in the coming generations as a way to greatly reduce the PC.

References

[1] Khan, M., Alhumaima, R. S. and Al-Raweshidy, H. S., "Reducing energy consumption by dynamic resource allocation in C-RAN," European Conference on Networks and Communications (EuCNC), 2015, pp. 169-174.

[2] Zhou, T., Huang, Y., Fan, L., Yang, L.: 'Load-aware user association with quality of service support in heterogeneous cellular networks', Commun. IET., 2015, **9**, (4), pp. 494–500.

[3] Yang, C., Chen, Z., Xia, B., Wang, J.: 'When ICN meets C-RAN for HetNets: an SDN approach', Commun. Mag. IEEE, 2015, **53**, (11), pp. 118–125.

[4] Basta, A., Kellerer, W., Hoffmann, M., Hoffmann, K., Schmidt, E. D.: 'A Virtual SDN-Enabled LTE EPC Architecture: A Case Study for S-/P-Gateways Functions', in Future Networks and Services (SDN4FNS) IEEE, 2013, pp. 1–7.

[5] Shea, R., Wang, H., Liu, J.: 'Power consumption of virtual machines with network transactions: Measurement and improvements', in IEEE INFOCOM 2014 - IEEE Conference on Computer Communications, 2014, pp. 1051–1059.

[6] Ricardo, L.: 'Evaluating the performance and power consumption of systems with virtual machines', in Cloud Computing Technology and Science (CloudCom), 2011 IEEE Third International Conference on, 2011, pp. 778–783.

[7] Aissioui, A., Ksentini, A., Gueroui, A. M., Taleb, T.: 'Toward Elastic Distributed SDN/NFV Controller for 5G Mobile Cloud Management Systems', Access IEEE, 2015, **3**, pp. 2055–2064.

[8] Desset, C., Debaillie, B., Giannini, V., et al.: 'Flexible power modeling of LTE base stations', in Wireless Communications and Networking Conference (WCNC), 2012 IEEE (2012), pp. 2858–2862.

[9] Liu, K., He, J., Ding, J., Zhu, Y., Liu, Z.: 'Base station power model and application for energy efficient LTE', in Communication Technology (ICCT), 2013 15th IEEE International Conference on, (2013), pp. 86–92.

[10] Auer, G., Giannini, V., Desset, C., et al.: 'How much energy is needed to run a wireless network?', Wirel. Commun. IEEE, 2011, **18**, (5), pp. 40–49.

[11] Holtkamp, H., Auer, G., Giannini, V., Haas, H.: A Parameterized Base Station Power Model, Commun. Lett. IEEE, 2013, **11**, (17), pp. 2033-2035.

[12] Alhumaima, R. S., Khan, M., Al-Raweshidy, H. S.: 'Component and parameterised power model for cloud radio access network', Commun. IET, 2016, **10**, (7), pp. 745–752.

[13] Alhumaima, R. S., Al-Raweshidy, H. S.: 'Evaluating the energy efficiency of software defined-based cloud radio access networks', Commun. IET, 2016, **10**, (8), pp. 987–994.

[14] Li, Y., Chen, M.: 'Software-Defined Network Function Virtualization: A Survey', Access IEEE, 2015, **3**, pp. 2542–2553.

[15] Huang, Q., Gao, F., Wang, R., Qi, Z.: 'Power Consumption of Virtual Machine Live Migration in Clouds', in Communications and Mobile Computing (CMC), 2011 Third International Conference on, (2011), pp. 122–125.

[16] Marcu, M., Tudor, D.: 'Power consumption measurements of virtual machines', in Applied Computational Intelligence and Informatics (SACI), 2011 6th IEEE International Symposium on, (2011), pp. 445–449.

[17] Stoess, J, Lang, C., Bellosa, F.: 'Energy Management for Hypervisor-Based Virtual Machines', in USENIX annual technical conference, (2007), pp. 1–14.

[18] Ye, L., Gniady, C., Hartman, J. H.: 'Energy-efficient memory management in virtual machine environments', in Green Computing Conference and Workshops (IGCC), (2011), pp. 1–8.

[19] Maurizio, P., Christian, E.: 'Network consolidation for virtualized servers', US Patent, **8**, 2011.

[20] Lauri, M., Brad, E.: 'The problem of power consumption in servers', Intel Corporation. Dr. Dobb's, 2009.

[21] Minjoong, K., Yoondeok, J., Jinseok, C., Moonju, P: 'A Simple Model for Estimating Power Consumption of a Multicore Server System', International Journal of Multimedia and Ubiquitous Engineering, 2014, **9**, (2), pp. 153–160.

[22] Wu, S., Zhou, L., Sun, H., Jin, H., Shi, X.: 'Poris: A Scheduler for Parallel Soft Real-Time Applications in Virtualized Environments', Transactions on Parallel and Distributed Systems IEEE, 2016, **27**, (3), pp. 841–854.

[23] Kim, C., Park, K. H.: 'Credit-Based Runtime Placement of Virtual Machines on a Single NUMA System for QoS of Data Access Performance', Transactions on Computers IEEE, 2015, **64**, (6), pp. 1633–1646.

INDEX

M

N

O

P

Q

R

S